T0186475

Innovations in GIS 5

Selected Papers from the
Fifth National Conference on
GIS Research UK (GISRUK)

Innovations in GIS 5

Selected Papers from the
Fifth National Conference on
GIS Research UK (GISRUK)

EDITED BY

STEVE CARVER

School of Geography
University of Leeds

UK   Taylor & Francis Ltd, One Gunpowder Square, London EC4A 3DE
USA Taylor & Francis Inc., 1900 Frost Road, Suite 101, Bristol, PA 19007

British Library Cataloguing-in-Publication Data

A catalogue record for this book is available from the British Library
ISBN 0 7484 0810 X

Library of Congress Cataloging-in-Publication Data are available

Cover design by Hybert Design & Type, Waltham St Lawrence, UK
Typeset in Times 10/12pt by Graphicraft Typesetters Ltd, Hong Kong
Printed in Great Britain by T.J. International Ltd, Padstow, UK

Contents

Foreword

I have never been asked to write a foreword before, and may never again once you read this! However, I wish to use this unique opportunity to stand back and offer a view of the GIS world, only part of which was represented at GISRUK 97. Maybe it will help you (the reader) to put things into perspective, or maybe it reflects a set of purely personal experiences, but looking around the world of GIS you can start to identify different types of GIS citizens. It is these that are the subject of this Foreword.

Some of the leading GIS figures in both academia and industry are what can be termed *strategic worriers*. They do not do GIS any more. They do not use GIS, if indeed many ever did, but they know all about it (or at least think they do). They do GIS by talking about it and, more importantly, by sitting on committees that organize the formal world of GIS. What they do well is to worry about key issues such as data copyright, standards for transfer protocols, policy issues and data access. What they do less well is to know about the research that is going on. They tend to be senior (i.e. old) within the GIS world, but very well connected at all the top levels. The 'gods' of the GIS hall of fame are their personal friends. This is the international élite of the GIS networkers, flying from one country to another. They must have massive Christmas card lists. This is the pinnacle of a life-time career in GIS. Wining and dining on a global scale cannot be at all bad, if that is what you like. Sadly these types are not much seen at GISRUK unless giving keynote addresses! More of them need to be lured here in the future to see for themselves what is happening at the grass-roots level of GIS research.

A very different category are the *niche finders*. These are flexibly minded GIS entrepreneurs who always are on the lookout for new business opportunities whatever and wherever they may be. Many of these business opportunities are mundane (e.g. cleaning up addresses) but can be highly profitable. They focus on solving specific and generic problems of global market relevancy. If you want to start your own small company and end up rich, then this is one way of doing it. Many of the ideas and innovations represented here probably originate from personal research at higher degree level. I wonder how many of the GISRUK presenters or poster-givers end up with stands at the AGI Annual Conference within the next five years? The niche finders provide a major technology transfer route and are to be much admired, because they personally transfer what may be initially fairly useless and esoteric research into useful products.

Then there are the *GIS vendors* themselves, whose senior executives carefully balance their perceptions of the needs of users with the performance of their equity on the stock markets. The vendors are highly competitive and are generally on a global convergence course, if only they realized it. In ten years time will there be any real difference between the various systems? Will GIS merely be an icon in Windows NT 12.0? Will users be locked into one proprietary system, or will there be transparent multi-system users with functionality scattered around all over the Internet. Probably the latter is what will happen in a highly invisible way, as meta (i.e. universal) GIS languages or interfaces are created than can sit on top of any and all existing systems. It is unlikely that these systems will do much that is very different from what is done today, but they may well do it differently and perhaps even more efficiently. The only GIS vendors seen at GISRUK are researchers who do not yet realize that one day they could perhaps become the GIS salesmen of the future. Remember that Jack Dangermond must have been at school once!

Another category of those present at GISRUK are termed the *appliers and improvers*. There is still time and opportunity for new entrants who come without the baggage of the past decade or so of GIS, and may thus be better able to devise the new system frameworks of the future. The GIS market place is now changing so rapidly that probably no vendor can anticipate likely developments more than three years ahead. One of the problems here is the popular belief in the commercial world that the best is the enemy of the good. Academic researchers are trained in a mindset that is dedicated to developing the best solutions; second best is just not good enough, because it *is* second best. Excellence and optimality is the name of the academic game. In a commercial context, however, best becomes irrelevant, because the good (a relative concept) is sold either as being the best (hence nothing better can exist without fear of embarrassment) or sufficiently close to the best (albeit undefined) for any practical purposes (they hope). This immediately isolates the world of research from the world of commerce. Entry for the academic is dependent on substituting good for best, on exchanging a fast approximation for a rigorous proof or a sophisticated optimal method, and of replacing a lengthy timescale spanning several weeks, months or years to one measured in hours or a few days (including weekends). Experience suggests that if you want to be an applier or improver (or a GIS innovator), you have to compromise some, but not all, of these academic virtues. There are then various technology transfer routes open to you: (1) publish and give it away; (2) try and sell it, but I bet it will have no value (the vendors are bound to tell you that your invention is only valuable when embedded within their system, and that the marginal added value it brings to the party is measurable but very small); (3) train people who can be hired by vendors and who are able to import your technologies by the back door; and (4) leave academia and do it all yourself. The last is the best, but it is hard work. Not many academics have it in them, as it requires exceptional research and managerial skills and considerable perseverance. GMAP Ltd is a glowing example of what can be done, given these skills.

Far less effort and risk is involved if you belong to a whole new generation of *GIS watchers and geopublishers*. Some are journalists who can easily make a good living out of describing the patterns and trends observed at conferences, exhibitions, and in the market place. In essence, the watchers watch others do GIS and report new developments, provide interviews, and generally try to make sense of it all. GIS publishing is a booming industry and it is nice to see 'the geographical' receiving so much press if only because it upsets the anti-GIS lobby even more!

This brings us to a related but more pernicious category – the *moaners*. They do not do GIS, they just moan about how it is or could be (mis)applied. At best they are the social conscience of the GIS world. They may claim to hold the moral high ground in terms of understanding the broader aspects of GIS ethics, morality, and GIS-related social (in)justice. It may be that they are jealous that GIS has left them behind. They may be sincerely concerned at the unfettered and unregulated explosion of GIS-based systems that are living embodiments of what they consider to be a wholly outmoded positivist approach which is actively harming people (and sometimes their ancestors' spirits) in very damaging theoretical ways that most GISers simply do not understand (or indeed want to understand). Some moaners are purely destructive (in another life they would probably be car wreckers in a scrapyard, but they choose university instead), while others try (or at least pretend) to be constructive. Of course, there is validity in some of these concerns, but too much is ill-founded: an exaggerated tornado in a teacup. Unfortunately, perhaps, there was too little of this at GISRUK. The critics need to be confronted and deconstructed – not excluded or ostracized. Sadly, it seems the GIS world is already segregated and you have to believe in GIS before you go to a GISRUK.

Next are the *GIS end-users*. There are probably well over a million GIS workers who use GIS technologies on a daily basis. These numbers are rapidly expanding on a global scale. Indeed GIS is a global business, albeit one dominated in the market place by large and rapidly growing (US-based) multinationals. Users merely want systems that work and which do what they want (whatever that is). Amazingly the end-users are very neglected. The initial users may only have had simple needs, but this may no longer be true. Unfortunately, there are few or no end-users here to offer their perspectives on the GIS research themes that matter to them. End-users do not do research and so may not find a research conference useful. Opportunities may need to be found to involve more of them.

Then there is an important category of *GIS starters*. It is this generation of young, vibrantly enthusiastic, dynamic, and energetic researchers that are needed and are expected to define the GIS systems of the future. Put simply, the GIS that exist today were conceived by a relatively small number of such young, vibrantly enthusiastic, dynamic and energetic researchers in the 1970s and early 1980s. The functionality of today's systems reflects very much the ideas, the vision, the enthusiasm and concepts of 20-odd years ago. Where is the next Terry Coppock or Mike Goodchild or Donna Peuquet? Certainly the technology is better now, the algorithms may have improved and the computer environments are totally different, but the functionality is essentially evolutionary. So where are the new ideas? Where are the visionaries of the new generation, able to define and lay the foundations for the systems of 2020 and beyond? There is concern that the GIS ideas lode may have been mined out. The new generations of GIS researchers are simultaneously brainwashed by existing systems (their long learning curves leave little time for anything else) and crippled (oops! I mean limited in what they can do) by the lack of programming, statistical, and mathematical skills. Perhaps, by failing adequately to train GIS researchers rather than users, we are committing the future of GIS to be more of the same rather than something new and different.

Other categories are harder to define, but a final one must be the *victims of GIS*. Victims, it seems, never present papers at conferences. Perhaps it is time they did. Of course this could be short-lived, as there is the prospect of a rapid future expansion of GIS-related litigation once the legal profession begins to understand enough of the technology to realise the opportunities open to their own form of legalized vulturism. Of course they would not view themselves in such a light, but the first thing they would do would be to silence the victims.

This is perhaps a good place to stop, as this Foreword may have become far too backward-looking for its own good. This GISRUK conference in Leeds clearly illustrates some of the vibrancy of the ongoing research. To make it big in GIS you need luck and good fortune as well as hard work and enthusiasm. Most of the GISRUK attendees have these latter properties in abundance. There can be no doubt that some of the stars of the future were in Leeds for a short while in April 1997. However, rather like a rock concert you have to be there to see and feel it, conference proceedings (no matter how well assembled as, indeed, this one is) only convey less than 5% of a GISRUK. So why not go and visit one yourself?

STAN OPENSHAW
School of Geography, University of Leeds
September 30th 1997

Preface

This book, representing volume 5 of a series, contains selected papers from the Fifth National Conference on GIS Research UK 1997 (GISRUK 97) which was held at the University of Leeds in April 1997. The 21 chapters published here are intended to be not only representative of the 80 or so papers and 20-plus posters presented at the conference, but also of some of the current research directions in GIS within the UK as a whole. To this end, the book is organized into five sections covering: spatial analysis; virtual GIS; artificial intelligence, spatial agents and fuzzy systems; space-time GIS; and GIS applications. Within these general headings the range of subjects covered by the collected chapters is remarkable. All are of exceptional quality.

One of the stated aims of the GISRUK conference is that it should provide a framework in which postgraduate students can see their work in a national context. The papers published here are, therefore, written not only by established researchers (the 'strategic worriers' of Openshaw's foreword) but by postgraduates just beginning their GIS research careers. These are the 'GIS starters' who are, in Openshaw's own words, the 'generation of young, vibrantly enthusiastic, dynamic, and energetic researchers that are . . . expected to define the GIS systems of the future'. GISRUK can be happy in the fact that it attracts a great many of this important category of GIS citizenry to its doors, and long may it continue to do so!

The conference itself followed the established GISRUK format; friendly and informal, with a variety of presentation styles and opportunities for discourse (learned and otherwise). The conference was enhanced by three excellent keynote speakers: Mike Goodchild from the National Centre for Geographic Information and Analysis; Mike Shiffer from the Michigan Institute of Technology; and Christiane Weber from University Louis Pasteur, Strasbourg. This year's conference also saw the return of the 'just-a-minute' sessions as a means of allowing poster-presenters time to introduce their work to the assembled conference audience.

This short preface provides me with the opportunity to thank all those involved with GISRUK 97 and especially those who made it all possible. The main sponsors of the conference were again the Association for Geographic Information (AGI), Taylor & Francis Ltd. and the Regional Research Laboratories Network (RRLNet). All these organizations are long-term sponsors of the GISRUK conference series and deserve particular thanks. May their support continue and further develop this special relationship. Many thanks go to the GISRUK Steering Committee and Local Organizing Committee members, in particular those people at the sharp end of making sure the conference ran smoothly. Particular thanks must go to Maureen Rosindale for her cool and efficient approach to essential bookings, to Jane Copeland for organizing registration and to Makis Alvanides for his many hours hard work compiling the conference proceedings. A special thanks must also go to all the postgraduates who gave up their own time to do all the odd jobs that I forgot about until the last minute. Thanks must finally also go to all the conference delegates, presenters and contributors to the book. Without them, GISRUK is nothing but a committee and an idea.

STEVE CARVER
University of Leeds, 1997

Contributors

Peter Atkinson
Department of Geography, University of Southampton, Highfield, Southampton, SO17 1BJ

Jonathan Baldwin
Geography Department, University of Leicester, University Road, Leicester, LE1 7RH
(JABDI@le.ac.uk)

A. Basden
IT Institute, University of Salford, Salford, M5 4WT
(kinds-request @mailbase.ac.uk)

Chris Bird
Silsoe College, Cranfield University, Cranfield, Bedfordshire, MK43 0AL

Allan Brimicombe
School of Surveying, University of East London, Longbridge Road, Dagenham, Essex, RM8 2AS

James Bullock
Department of Geography, University of Southampton, Highfield, Southampton, SO17 1BJ

Steve Carver
School of Geography, University of Leeds, Leeds, LS2 9JT

Spencer Chainey
London Borough of Hackney, Policy and Research – Directorate of Housing, Christopher Addison House, 72 Wilton Way, London E8 1BJ

Christophe Claramunt
The Nottingham Trent University, Department of Computing and Geography, Burton Street, Nottingham, NG1 4BU

K. Cole
MIDAS, Manchester Computing Centre, University of Manchester, Manchester, Oxford Road, Manchester M13 9PL
(kinds-request@mailbase.ac.uk)

Steven H. Cousins
IERC, Cranfield University, Cranfield, Bedfordshire, MK43 0AL

Noemi De La Ville
IERC, Cranfield University, Cranfield, Bedfordshire, MK43 0AL

Zhiqiang Feng
Department of Geography, Lancaster University, Lancaster, LA1 4YB
(z.feng@lancaster.ac.uk)

Robin Flowerdew
Department of Geography, Lancaster University, Lancaster, LA1 4YB
(r.flowerdew@lancaster.ac.uk)

Roger Glover
Department of Geography, University of Nottingham, University Park, Nottingham, NG7
2RD

Michael F. Goodchild
NCGIA, University of California, Santa Barbara, CA 93106-4060, USA
(good@ncgia.ucsb.edu)

Ian Gregory
Department of Geography, Queen Mary and Westfield College, University of London,
Mile End Road, London, E1 4NS

Cédric Grueau
Centro Nacional de Informação Geográfica, Rua Braamcamp 82, 5° Esq, 1250 Lisboa,
Portugal

A. Heptinstall
Department of Environmental and Geographical Sciences, Manchester Metropolitan University, Manchester, M1 5GD
(kinds-request@mailbase.ac.uk)

Heike Hofmann
Mathematics Institute, University of Augsburg, 86135 Augsburg, Germany

M.N. Islam
IT Institute, University of Salford, Salford M5 4WT
(kinds-request@mailbase.ac.uk)

Zarine Kemp
Computing Laboratory, University of Kent at Canterbury, Canterbury, Kent, CT2 7NF

K. Kitmitto
MIDAS, Manchester Computing Centre, University of Manchester, Manchester, M13
9PL
(kinds-request@mailbase.ac.uk)

Roman M. Krzanowski
Department of Geography, Birkbeck College, University of London, 7–15 Gresse Street,
London, W1P 2LL

Howard Lee
Computing Laboratory, University of Kent at Canterbury, Canterbury, Kent, CT2 7NF

C.S. Li
MIDAS, Manchester Computing Centre, University of Manchester, Manchester, M13
9PL
(kinds-request@mailbase.ac.uk)

Jenny Lingham
Department of Computer Science and Geography, Keele University, Staffordshire, ST5
5BG

Paul Longley
Department of Geography, University of Bristol, University Road, Bristol, BS8 1SS

David Martin
Department of Geography, University of Southampton, Southampton, SO17 1BJ
(D.J. Martin@soton.ac.uk)

Keith Mason
Department of Environmental and Social Sciences, Keele University, Staffordshire, ST5
5BG

Victor Mesev
Department of Geography, University of Bristol, University Road, Bristol, BS8 1SS
(t.v.mesev@bris.ac.uk)

Adrian Moss
Department of Environmental and Geographical Sciences, Manchester Metropolitan University, Manchester, M1 5GD
(kinds-request@mailbase.ac.uk)

Nuno Neves
Centro Nacional de Informação Geográfica, Rua Braamcamp 82, 5° Esq, 1250 Lisboa,
Portugal

Abigail Nolan
Department of Geography, University of Southampton, Highfield, Southampton, SO17
1BJ
(amn2@soton.ac.uk)

Stan Openshaw
School of Geography, University of Leeds, Leeds, LS2 9JT

Christine Parent
Swiss Federal Institute of Technology, DI-LBD, Lausanne, CH-1015, Switzerland

Scott Parsley
Department of Geomatics, Bedson Building, University of Newcastle upon Tyne, Newcastle upon Tyne, NE1 7RU
(scott. Parsley@ncl.ac.uk)

Jim R. Petch
Department of Environmental and Geographical Sciences, Manchester Metropolitan University, Manchester, M1 5GD
(kinds-request@mailbase.ac.uk)

Gary Priestnall
Department of Geography, University of Nottingham, University Park, Nottingham, NG7 2RD
(gary.priestnall@nottingham.ac.uk)

Jonathan Raper
Department of Geography, Birkbeck College, University of London, 7–15 Gresse Street, London, W1P 2LL

Armanda Rodrigues
Centro Nacional de Informação Geográfica, Rua Braamcamp 82, 5° Esq, 1250 Lisboa, Portugal
(armanda@cnig.pt)

Linda See
School of Geography, University of Leeds, Leeds, LS2 9JT
(L.See@geog.leeds.ac.uk)

Humphrey R. Southall
Department of Geography, Queen Mary and Westfield College, University of London, Mile End Road, London, E1 4NS

Neil Stuart
University of Edinbrugh, Department of Geography, Drummond Street, Edinburgh, EH8 9XP

George Taylor
Department of Geomatics, Bedson Building, University of Newcastle upon Tyne, Newcastle upon Tyne, NE1 7RU
(George. Taylor@ncl.ac.uk)

Marius Thériault
Laval University, Department of Geography, Planning and Development Research Centre, Quebec, GIK 7P4, Canada

Anthony Unwin
Mathematics Institute, University of Augsburg, 86135 Augsburg, Germany
(unwin@uni-augsburg.de)

Christiane Weber
IMAGE et VILLE (CNRS), University Louis Pasteur, 3 rue de l'Argonne, 67 000 Strasbourg, France
(chris@lorraine.u-strusbg.fr)

Michael Worboys
Department of Computer Science and Geography, Keele University, Staffordshire, ST5 5BG
(michael@cs.keele.ac.uk)

Y.J. Yip
IT Institute, University of Salford, Salford M5 4WT
(kinds-request@mailbase.ac.uk)

GISRUK Committees

GISRUK NATIONAL STEERING COMMITTEE

Steve Carver	University of Leeds, UK
Jane Drummond	Glasgow University, UK
Peter Fisher	University of Leicester, UK
Bruce Gittings	University of Edinburgh, UK
Peter Hall	York University, UK
Gary Higgs	Cardiff University, UK
Zarine Kemp	University of Kent, UK
David Kidner	University of Glamorgan, UK
David Martin	University of Southampton, UK
David Parker	University of Newcastle upon Tyne, UK

GISRUK '97 LOCAL ORGANIZING COMMITTEE

Steve Carver
Stan Openshaw
Makis Alvanides
Heather Eyre
James Macgill
Maureen Rosindale
Jane Copeland

GISRUK '97 SPONSORS

Association for Geographic Information (AGI)
Regional Research Laboratories Network (RRLnet)
Taylor & Francis Ltd.

Introduction

I

As we move towards the new millennium and the GIS bandwagon trundles along at ever increasing speed and momentum, the annual GISRUK conference gives those of us with GIS research interests at heart a short breathing space in which to pause, listen, observe, discuss and otherwise reflect on where the subject is heading. It provides an excellent opportunity to catch up on what the rest of the UK GIS research community is up to, while presentations from guest speakers and other non-UK authors act to keep us abreast of what is happening elsewhere in the world.

One of the best characteristics of GISRUK to my mind is that is it almost wholly devoid of commercial hype and the hard sell that we see at many other GIS conferences and exhibitions. In this manner GISRUK is a breath of fresh air, for as its name implies, GIS *Research* UK is entirely reserch led. Sure, GIS researchers often rely on commercial systems, but there is an under-current within GIS research in its broadest sense that has been forced to move away from off-the-shelf GIS packages because of their lack of advanced functionality. Spatial analysis is a case in point. Here we see many authors using not GIS, but statistical packages, graphics packages, algorithm libraries and custom software to achieve their desired aims. As with other techniques-based GIS research, it is to be hoped that this type of research will feed into the development of existing commercial systems or spawn new add-ons. The GIS packages we all know and love today, as Openshaw recounts in his foreword, are the product of research work done in the 60's and 70's. Research being done today will ultimately become the functionality of tomorrow.

Another observation which is particularly pertinent to this book and others in the series, is the sheer volume of GIS texts, magazines and journals available today. Our bookshelves are already groaning under their combined weight. Can they take any more? To a certain extent this profusion of written material is an indication of the health of the subject and its broad appeal. As GIS matures then the amount of published material available mushrooms as publishers realise the demand and react with writers like myself to identify gaps on the shelf, target niches and generate the supply. The sobering thought from my point of view is that I no-longer have what I feel to be a detailed enough knowledge of the whole GIS discipline, rather I inevitably find myself knowing more and more about less and less as I specialise in particular fields of this burgeoning subject area. I am sure many readers feel the same. In this very respect, GIS texts like this one fill an important niche in a nearly saturated market. They are a snap-shot in time that provide the reader with a sample of the current state of play regarding research being done within the UK GIS community. As the series accumulates, they will surely provide an excellent potted research history that chronicle developments within the discipline.

The contents of this particular volume reflect certain emergent fields in GIS research and new developments in some existing ones. The book is divided into five parts, each containing chapters related to as common a theme as is possible given the diverse nature of GIS in the late 90's. Part One covers that old favourite **Spatial Analysis** which still produces a constant stream of high quality papers. Part Two records developments in a new field, that of **Virtual GIS** and associated GIS applications on the World Wide Web. Part Three deals with **Artificial Intelligence, Spatial Agents and Fuzzy Systems**; a rather long title intended to catch a number of innovative works in the general area. Part Four concerns **Space-time GIS** and includes a number of papers with a temporal GIS theme. Finally, Part Five provides room for a number of papers with an **Applications** theme. One the one hand this last section is a catch-all for selected papers that did not really fit into any of the other themes and on the other hand it is intended to allow the inclusion of some purely applied works where GIS is seen more as a tool rather than the subject of the research itself. One view is the GISRUK should be about GIS research, not its application. However, it is recognised here that, to a certain extent, the need for GIS research is led by the needs generated by new fields of application and as such purely applied work should not be excluded.

SPATIAL ANALYSIS

The four papers in this section deal with subjects ranging from the analysis of flow data to exploratory spatial data analysis.

Spencer Chainey's and **Neil Stuart's** paper revisits that age-old problem of spatial interpolation from point observations. They develop an alternative technique to those already available within proprietary GIS packages based on stochastic simulation. One of the main problems with standard techniques of spatial interpolation is that of the uncertainty surrounding the accuracy of the resulting surface. The advantage of using a stochastic simulation approach is that it offers the user a more realistic measure of the uncertainty in the interpolation across the mapped area. Such a tool would not only give users better interpolations but also provide better estimates of uncertainty on which to base decisions made.

The paper by **Roman Krzanowski** and **Jonathan Raper** develops a new spatial evolutionary algorithm for the modelling of spatial problems based on genetic algorithms. This is presented and tested on a range of different planar coverage problems and example data sets to demonstrate their response to the key variable of population size and show their learning probabilities. The results demonstrate the advantages of evolutionary algorithms over non-evolutionary heuristics seen hitherto in this field.

The paper by **Mike Worboys, Keith Mason** and **Jenny Lingham** describes the development of an approach to spatial analysis in the setting of a non-Euclidean space. This is interesting because by far the greater majority of GIS functionality assumes that entities, observations, etc. are located and referenced within Euclidean space. Patterns of migrant flows are given as an example where the Euclidean model may not be the most appropriate (due to changes in social structure, planning and transportation) and is used to demonstrate a new method for the treatment of non-Euclidean proximity relationships. Potential applications of these techniques abound in societies where the effects of geographical distance are either foreshortened or removed altogether by technological advances in transport and communications.

Anthony Unwin's and **Heike Hofman's** paper closes this section with their description of a new interactive graphics tool (MANET) developed originally as a means of coping with data sets with missing values. This package is applied here for the exploratory spatial analysis of demographic and social data concerning 437 counties in six states of the American Mid-West. This paper represents another useful example of where advanced spatial analysis is carried out external to proprietary GIS systems albeit with recourse to GIS principles as appropriate.

VIRTUAL GIS

Three chapters have been selected under this title to represent new directions in the development of web-based GIS applications.

Mike Goodchild's paper on current work in the USA in developing virtual geolibraries was one of the invited contributions to GISRUK '97. The paper describes geolibraries as internet-based containers of geographically referenced information objects and goes on to identify their essential components and current technological limitations. Existing geolibrary-type projects such as the Alexandria Digital Library are reviewed. The paper, rather inevitably for such a forward looking piece, closes with a lengthy list of research questions focusing on copyright, cost, organisational issues, standards, data models and data quality.

Following directly on from this work is the paper by **Adrian Moss** and co-workers concerning the KINDS (Knowledge-based Interface to National Data Sets) Project. This project identifies the main limitations to uptake and usage of spatial data as lack of awareness of their existence and poor data set accessibility. The KINDS Project aims to increase the user base of national data sets through the development of a common web-based interface to the search, browse and map requirements of users. Encouraging results showing ease of access across a range of users (from novice to expert) are reported.

The final paper in the section by **Scott Parsley** and **George Taylor** shows how the principles of virtual reality modelling can be applied to the accurate measurement, re-cording and visualisation of three dimensional data of the urban environment. The paper shows how virtual GIS worlds of historic city landscapes can be constructed from old photographs and used to assist in archaeological measurements and reconstruction.

ARTIFICIAL INTELLIGENCE, SPATIAL AGENTS AND FUZZY SYSTEMS

A total of five chapters in this section deal with various aspects of AI and the application of fuzzy models to spatial problems.

The first paper in this section was presented by another of the invited speakers, **Christiane Weber**. This deals with multi-agent systems (MAS) approaches to the man-agement of geographical problems. The example used is that of (i) the propagation of groundwater pollution and (ii) the decision process resulting from different actors dealing with the ensuing crisis. Simulations of pollution incidents are used to test management responses and develop adaptations to these.

The paper by **Linda See** and **Stan Openshaw** details some recent experiments in building fuzzy models of spatial data. Two examples of fuzzy model building in a GIS data rich environment are described. The first of these looks at the problems of creating fuzzy spatial interaction models and their potential for spatial decision support systems, while the second considers the development of a simple model of employment. Results show that there is plenty of potential for fuzzy modelling within GIS especially given the increasing complexity of spatial models and their applications.

Armanda Rodrigues, Cedric Grueau, Jonathan Raper and **Nuno Neves** present a paper on the incorporation of spatial reasoning via multi-agent systems within spatial decision support systems. The example of environmental planning in land use management is given and MAS shown to improve spatial decision making within this context. The paper complements the ideas and work presented by Christiane Weber.

Zhiqiang Feng and **Robin Flowerdew** present a further paper utilising fuzzy modelling techniques. This paper discusses using fuzzy clustering methods to classify census data and set up fuzzy geodemographic systems. Feng and Flowerdew show how fuzzy membership values derived from enumeration district data can be used to undertake detailed customer profiling.

The final paper in this section by **Christophe Claramunt, Marius Thérialt** and **Christine Parent** deals with spatio-temporal models for representing the evolution of two dimensional spatial entities. In their paper they propose a qualitative reasoning approach based on inferences of measure time (i.e. chronology) and historical time (i.e. ordering) regarding geometrical and topological representations of space. As an intelligent approach to temporal reasoning in GIS this papers sits on the boundary of this and the following section on Space-time GIS.

SPACE-TIME GIS

This section contains three chapters that deal with developments in temporal GIS research.

The first paper in this section has been written by **Alan Brimicombe** and concerns the development of a fuzzy co-ordinate system for locational uncertainty in space and time. Again, this is a paper that would sit equally well in the adjacent section within the ordering of the book. Here Brimicombe uses fuzzy numbers to construct a fuzzy co-ordinate system for use in two, three and four dimensions which may be incorporated within traditional GIS data structures.

Following on from this the paper by **Howard Lee** and **Zarine Kemp** describes a framework that may be used to support analyses with complex space-time characteristics. The research is part of an initiative to manage marine fisheries; an area where the ability to handle dynamic spatio-temporal data is advantageous. The authors use an analytical abstraction layer (AAL) to better manage spatiotemporal data and search for patterns.

Gary Priestnall's and **Roger Glover's** paper proposes a particular approach to automating the detection of land use change using existing vector databases to guide object-recognition from very high resolution remotely sensed imagery. Mis-matches between the vector outlines and the raster imagery can be used to identify changes and new features in imagery taken since the vector data was compiled. Such a tool has obvious applications in detecting change but could also be used to automate the updating of vector databases over time.

APPLICATIONS

As already suggested, this final section in the book may be regarded as a catch-all for papers that do not fit readily into any of the above sections or as a vehicle for purely applied GIS research. A total of 5 papers are included, covering a variety of subject areas.

The first paper in this section is by **Jonathan Baldwin** and shows how GIS may be used to support landscape planning and assessment. Viewshed analyses are used together with results from a questionnaire to determine the physical characteristics of the landscape that appeal to the viewer. These include horizons, viewshadows (unseen areas) and viewshapes. The applied aim of the paper is to enable landscape planners and managers to manage the damage to popular viewpoints and suggest alternatives as a means of reducing visitor pressure.

The second paper describes a habitat suitability analysis for the Grey Wolf in North America. This paper by **Noemi De La Ville, Steven Cousins** and **Chris Bird** utilises GIS and logistic regression to outline potential areas for conservation efforts. Habitat suitability is based on predictions of habitat quality using nine factors in a logistic regression model. The aim of the paper is to identify minimum sized areas of high suitability for the re-introudction of viable wolf populations. This is a topical suject in wildlife management as the US Fish and Wildlife Service attempt to re-introudce wolf packs into areas of Wyoming and central Idaho.

David Martin's paper addresses short comings in the design of recent censuses. Martin criticises the 1991 Census for not taking GIS into account and strongly argues the case that the 2001 Census be designed by and for GIS. The paper goes on to outline a prototype GIS for the automated creation of unit postcode building blocks and optimised output areas for the 2001 Census. The solution offered is based around merged Thiessen polygons around ADDRESS-POINT labels within constraining boundary sets.

A historical GIS database is described in the paper by **Ian Gregory** and **Humphrey Southall**. The period covered is from the mid nineteenth century to 1973 when digital boundaries first became available. The database developed is being used for historical GIS-based research into the socio-economic structure of Britain as it changed through time and space over the past 150 years.

The paper by **Victor Mesev** and **Paul Longley** attempts a preliminary investigation of the relationship between the physical and social functioning of urban settlements based on a case study of Norwich. Remote sensing data and census boundary data are used to map regional density gradients. It is shown how high resolution digital remotely sensed imagery can profitably enhance interpretations of urban geography based on traditional techniques using administrative geography.

The final paper in this section and the book is by **Abigail Nolan, Peter Atkinson** and **James Bullock** and demonstrates the use of GIS to examine change within the lowland heaths of Dorset between 1978 and 1987. An object-oriented approach is used to examine heath land dynamics at the patch level to derive information of use to heath land managers.

PART ONE

Spatial Analysis

Stochastic simulation: an alternative interpolation technique for digital geographic information

SPENCER CHAINEY AND NEIL STUART

1.1 INTRODUCTION

Few would disagree that good-quality initial data is needed to successfully produce an effective final solution from a geographical information system (GIS). However, while we often think of the poor quality of spatial data as being associated with inaccurate observations or those made over too short a period of time, a more basic problem with much geographical data is the sparseness of locations at which samples are taken. Possible reasons for this may include the inaccessibility of some sites and the prohibitive costs that may be involved in carrying out a full survey. If, as on many occasions, we then wish to create a map or other realization showing the full spatial extent of a variable (based on the limited observations available), we may use one of a number of spatial interpolation methods. Spatial interpolation allows us to estimate the value of a property at a relatively large number of unsampled locations as a function of the relatively few observed point values.

However, for all locations where a value has been estimated, there will be an inherently element of uncertainty about how closely the estimated value approaches the correct value that would actually exist at this location if an observation were to be taken. This dimension of error is often 'hidden' from the user who may instead be caught in the 'false lure about the attractive high-quality cartographic products that cartographers, and now computer graphics specialists, provide for their colleagues in environmental survey and resource analysis' (Burrough, 1986, p. 103). For example, Figure 1.1(a) shows 10 sample points distributed randomly across an area; Figure 1.1(b) shows a continuous representation of this data. But how accurate is (b)? Furthermore, would the accuracy differ if a different interpolation technique were used? And how different might the continuous surface representation look if the initial observation points were at different locations? The ground 'truth' is never known in a real situation – that is precisely why one needs to interpolate – yet questions about the relative properties of interpolators do need to be

3

(a)   (b)

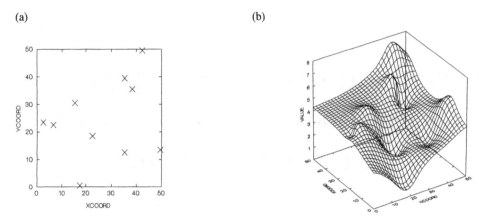

Figure 1.1   (a) Random distribution of 10 sample points over an area. (b) The continuous coverage that results by using the smoothing interpolation function from the SYSTAT computer package.

resolved as a growing number of users of GIS, including environmental modellers, rely implicitly on these methods in the preparation of their models and results. (Burrough *et al.*, 1996; Goodchild *et al.*, 1993).

A helpful early review of the issues involved in selecting an interpolator to match both the requirements of a project and the limitations of the data is given by Lam (1983), and there is also a useful comparison of different spatial interpolation methods in Burrough's 1986 text. However, to find answers to these questions the geographical information science community has in recent years had to look mainly to applied work in geostatistics. In soil science and engineering there has been research into the relative accuracy of different methods of spatial interpolation for different data sets (e.g. Laslett, 1994; Laslett and McBratney, 1990; Voltz and Webster, 1990; Dubrule, 1984 and 1983), while more basic statistical research concerning the derivation of the uncertainty which should be attached to estimates produced by analytical methods, such as interpolation, has been provided by, amongst others, Oliver and Webster (1990), Englund (1993) and Journel (1996).

One strand of research to which the GIS community has actively contributed is regarding the visualization of uncertainty in geographical data sets. There have been several papers, including a special issue of the journal *Cartographica*, concerned with creating new ways to make the usually hidden dimension of error more visible in output products (Buttenfeld, 1993; Fisher, 1994; van der Wel *et al.*, 1994). This reflects a growing recognition that, for any data set upon which decisions are to be made, the user needs an explicit statement about the reliability of the data across the area which is being mapped or analyzed. This is particularly vital if the data are in fact estimations derived from limited field observations using a method of interpolation (see for example, MacEachren *et al.*, 1993).

This study aims to contribute to the debate about the potential of different interpolators for GIS users in two main ways;

1   First, we conduct a practical experiment to evaluate the flexibility and the accuracy of the stochastic simulation method for interpolating different types and arrangements of geographical data. We compare our results for stochastic simulation with those obtained from two other interpolation methods, kriging and Thiessen polygons, which

are commonly found in GIS. This comparison should help to reveal the strengths and limits of the method when it is applied to a range of geographical data sets.

2 Secondly, by analyzing the composite results produced over a series of stochastic simulations for any one data set, we develop a measure of the uncertainty of the interpolated estimates. By comparing these maps of interpolator uncertainty with corresponding maps of true error (which can be calculated in this experimental situation), we demonstrate a way of mapping the uncertainty of an interpolator that consistently matches well with the distribution of true error. This provides the user with a better impression of the reliability of values throughout an interpolated map than previously used measures such as, for example, kriging variance.

1.2 WHY CONSIDER STOCHASTIC SIMULATION FOR INTERPOLATION IN GIS?

Stochastic simulation allows the drawing of different, equally probable, high-resolution models of the spatial distribution of an attribute under study (Deutsch and Journel, 1992). The differences between these realizations can then be used to measure the uncertainty in the estimations. Although stochastic simulation has been used in many geological applications, including petroleum reservoir research (Dubrule, 1989; Haldorsen and Damsleth, 1990; Journel and Alabert, 1990) and mining (Journel, 1986), there are few reports of its application by the GIS community. Awareness of the method and its potential benefits for GIS is, however, increasing (see Srivastava, 1996; Journel, 1996; Atkinson and Kelly, 1996; Englund, 1993) and it seems likely that stochastic simulation may be considered for inclusion in the analytical toolboxes of GIS in the future.

Since it would be impractical to compare stochastic simulation against all the interpolators found within GIS at present, in this initial evaluation the interpolation methods that were selected to provide a comparison were Thiessen polygons and kriging. Interpolation using Thiessen polygons was selected because of its widespread use by the GIS community. Thiessen polygons produce a map by creating an 'area of influence' around the sample points. The method is quick to implement and is still frequently used in hydrological and environmental modelling. For example, in a project to estimate spatial patterns of crop yield, Carbone et al. (1996) used Thiessen polygons to create a GIS data layer of rainfall from rain gauge measurements. However, this method, which was initially a manual technique, is argued to be of limited value for accurately estimating the full coverage of an attribute from sample point data in an era of powerful digital analysis. Beyond the need for compatibility with traditional hydrological analysis, perhaps the strongest case for the continued use of Thiessen polygons as a general interpolation method is for use on categorical data, where most other methods are inappropriate.

Kriging was selected because of the method's reported high accuracy in estimating values at unsampled locations (Oliver and Webster, 1990; Laslett, 1994; Isaaks and Srivastava, 1989; Dubrule, 1983 and 1984; Voltz and Webster, 1990). An additionally attractive aspect of kriging has been the claim that an estimate in the error of the interpolation is produced as a by-product of the method. This quantity is the kriging variance. Concerns over using this quantity as a measure of uncertainty are, however, increasing. Deutsch and Journel (1992) note that the kriging variance is independent of the values at the sample data points. Instead, the kriging variance depends only on the geometrical arrangement of the sample data points – being highest in areas remotest from any data points. This is not the same as what most people would refer to as an estimate of 'true error', i.e. where the error is the difference between the value estimated by the interpolator and a true value which would be obtained by making an independent observation at the location. In general, it is suggested that, while kriging is an effective interpolator, the kriging variance gives little clue as to where, within a kriged map, estimates may be considered to be more or less reliable. Yet it is precisely such a measure of interpolator uncertainty (or conversely, map reliability) that is required to help interpret, make decisions, or parameterize models from such data with confidence.

These comments are not intended to suggest that either Thiessen polygons or kriging are inappropriate methods, but merely to demonstrate that each method has its strengths and weaknesses and also, perhaps, a time when it is in fashion. As the limitations of one method are discovered, research proceeds elsewhere. Partly because of observed limitations with kriging, the GIS research community has begun to evaluate the potential offered by a variety of other methods. One of these methods is stochastic simulation, known also as stochastic imaging (see Journel, 1996; Srivastava, 1996; Myers, 1996; Gotway and Rutherford, 1996; Englund, 1993; Deutsch and Jounel, 1992; Rossi et al., 1993; Bierkens and Burrough, 1993; Stoyan et al., 1987). Stochastic simulation rests on some of the same geostatistical assumptions and techniques as kriging, including the concepts of random variables and spatial dependency. How it differs to other interpolation techniques is that, rather than creating a single 'optimal' realization from point observations, the technique creates many different realizations, each of which is an equally 'good' representation of the reality. One major benefit of this simulation approach is that the differences between these realizations can then help to develop a numerical and visual model of the uncertainty of the estimations. A further appeal of stochastic simulation algorithms to the GIS community is the prospect that the method may offer a higher level of versatility than traditional interpolation techniques. This versatility includes:

- Producing an interpolated surface for a wider variety of geographical data sets (i.e. data sets that are either continuous, partially discontinuous, categorical or have distributions that fit some geometric shape).

- Reproducing the spatial variability in these data sets (i.e. accounting for both local spatial variability of a property and honouring the global statistics of the phenomena under study).

- Interpolating data whether the classifications are required for hard, Boolean sets or soft, fuzzy sets (Deutsch and Journel, 1992).

1.3 APPROACH AND AIMS

Whilst a number of researchers in spatial statistics have recently explored the theoretical advantages of stochastic simulation, few studies have explored the utility of the technique for real-world geographic data. We recognise that most analysts using GIS in operational applications do not have the time to compare many combinations of different interpolation methods and sampling schemes. The choice of interpolator is often made on the basis of familiarity or practical convenience. By providing some comparative results for common data types and sampling arrangements, we aim to address, at least partly, this need for practical advice on the likely effectiveness of using different interpolation techniques.

Through a series of controlled experiments on a variety of geographical reference data sets, we aim to assist the user to a clearer understanding of the relative accuracy of stochastic simulation, compared to other interpolation techniques commonly found in the GIS toolbox (namely, ordinary kriging and interpolation by Thiessen polygons). Specifically, we assess the relative accuracy of each interpolation method under different conditions of:

- The underlying nature of spatial variability (different kinds of geographic phenomena).
- The amount of sample data available for interpolation.
- The geometrical arrangement, or spatial distribution of data points within the region where interpolation is required.

As well as understanding the relative accuracy, those taking decisions or wishing to run models on the results of an interpolation need to know the likely uncertainty in the estimates produced. For the same reference geographic data sets we evaluate how a stochastic simulation method may provide a more realistic representation of the pattern and magnitude of the uncertainty of the estimations, compared with the more commonly used kriging variance.

The next section explains how the controlled experiments were designed and how the different interpolations were created and compared.

1.4 METHODOLOGY

1.4.1 Sample data sets for interpolation

Three influences on the accuracy of an interpolation technique are:

(a) The geographical nature of the attribute under study

Geographical phenomena vary in the way they exist in the real world. In general, geographical data sets can be continuous (e.g. elevation), categorical (e.g. land cover), or 'spatially variable', where changes in the attribute values occur quite abruptly, rather than smoothly. A spatially variable data set displays spatial dependence, but is also heterogeneous, such as rock porosity (influenced by fissures). Three data sets were therefore selected to represent the properties of these types of geographical data (see Figure 1.2). For the convenience of data format, reference data size, and accessibility for other researchers doing comparative work, a continuous data set and a spatially variable data set provided as experimental examples with the GSLIB software were used. While neither was of a particular geographic region, it was argued that the visual and numerical nature of the data sets was more important for this study than the geographic area from which they were taken. The categorical reference data set selected was of land cover for an area in Edinburgh, Scotland. Six types of land cover (built, grass, wood, shrub, water, and several unclassified pixels) are displayed in the coverage (Figure 1.2(b)). This coverage was produced from a classification of Landsat Thematic Mapper imagery (Zhang, 1996).

For evaluating accuracy, each reference data set selected was assumed to be the 'true' representation of the phenomena to be interpolated. It is acknowledged that errors in sample observations can contribute significantly to producing an interpolation that is only a poor reflection of the true distribution of an attribute in the real world. For the purpose of this study, however, it is the interpolation techniques themselves that are to be evaluated; thus errors in the original data were assumed not to exist.

(a) Map of continuous reference coverage

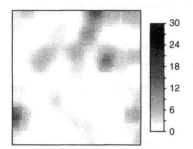

(b) Map of categorical reference coverage

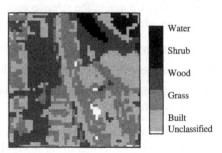

(c) Map of spatially variable reference coverage

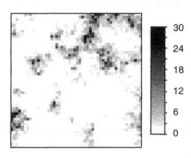

Figure 1.2   Reference data sets.

(b) Sample size

As the number of sample data points available increases, the relative accuracy of different interpolation techniques may vary. To allow for the effect that sample size may have upon interpolator accuracy, a number of different sample sizes were selected. These sample sizes (see Table 1.1) were deemed representative of the sizes of sample data typically available to GIS analysts.

(c) Sample distribution

Several studies where geographical data sets have been interpolated have concluded that the geometric arrangement of the data points throughout the study area can also strongly influence the accuracy of results obtained (for example, see Laslett, 1994). In this project, two of the main methods of sampling geographic data were investigated and compared (random distribution and systematic distribution). For this effect of sample distribution to be studied independently of sample size, interpolations were made for each of the sample sizes in Table 1.1 using both a random, then a systematic sampling distribution.

1.4.2   Measures of spatial variability and variogram modelling

Measures of spatial variability are central to any geostatistical study (Deutsch and Journel, 1992). The experimental variogram (or similarly used, experimental semivariogram) has traditionally been the method adopted to extract the necessary parameters required to

Table 1.1 Selected sample sizes and the percentage each represents of the whole coverage (50 × 50 pixels).

Sample size (Number of data cells)	Percentage of whole coverage
25	1
49	1.96
100	4
225	9
289	11.56

construct a model of spatial variability (the variogram model). The parameters extracted that collectively characterize the spatial variation of the data set under study are the nugget, sill, range, anisotropy, and type of variogram mathematical model fitted to the data (i.e. Gaussian, exponential, or spherical). Because of the complexities involved in modelling anisotropy, each data set to be tested was assumed to be isotropic in its distribution. Sample data sets were thoroughly explored in several directions to confirm that no strong anisotropic effects were present in the reference data sets to be used. This allowed a more careful and confident extraction of the parameters required for variogram modelling. A fuller description of this preparatory analysis is contained in Chainey (1996), and is based upon standard methods described by Isaaks and Srivastava (1989), and Webster and Oliver (1990).

1.4.3 Software

The Geostatistical Software Library (GSLIB) (Deutsch and Journel, 1992) was used for kriging and stochastic simulation. The specific algorithms utilized were:

- *Kriging*. Ordinary kriging algorithm for continuous and spatially variable data.
- *Stochastic simulation*. Gaussian simulation for continuous and spatially variable data, and categorical (probability density function) simulation for categorical data.

Reviews of these and other geostatistical interpolation algorithms can be found in Deutsch and Journel (1992), Varekamp *et al.* (1996), and Myers (1996). Ordinary kriging cannot, however, be applied to categorical data sets. Indicator kriging could have been applied, but was beyond the scope of this initial evaluation.

Of note also is the fact that in GSLIB, kriging and simulation algorithms work using covariance values. Thus, once a variogram has been modelled in the usual way, the software converts the estimated variogram model parameters into equivalent covariance models.

Thiessen polygons for each sample data set were created using a Turbo Pascal program developed by the authors, following a standard method for their construction as reported by Burrough (1986) and Ward and Robinson (1990).

1.4.4 Creating stochastic simulations

Stochastic simulation requires the same inputs as kriging – sample data and a variogram model (Englund, 1993). For continuous and spatially variable data sets it proceeds, however, by (see also Figure 1.3):

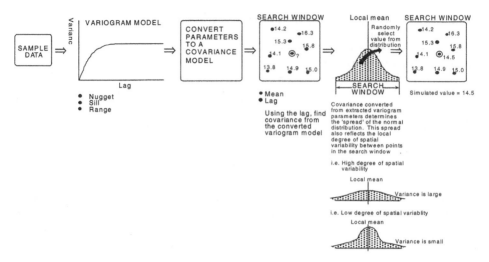

Based on the values of the sample data and their local variability, a simulated value of 14.5 is returned for the unsampled location

Figure 1.3 Gaussian simulation with GSLIB.

1   Choosing at random an unsampled location.

2   Within a defined search window determining the mean of the sample values and the lag between the conditional sample data and the unsampled location. This is to assess local spatial dependence between sample points and the unsampled location.

3   From the converted variance to covariance matrix, looking up the corresponding covariance for the computed lag.

4   Using the mean and covariance information, creating a frequency distribution describing the values within the search window.

5   Randomly selecting a value for the point from the distribution of the sample data within the search window.

The process then proceeds by following a random path to the next unsampled location, repeating steps 1–5 until all points have been simulated. Different equiprobable realizations are produced by different random paths for each simulation.

Rarely, however, are geographical data sets fully symmetrically Gaussian in their distribution. It is noted that by performing a normal score transform on all the sample data sets, simulation accuracy could potentially be improved. For this initial evaluation however, the histogram produced from each set of sample data was the accepted distribution.

For categorical simulations the method is slightly different. The global statistics of the data (i.e. each category's prior probability based on its relative frequency in the sample data) are initially calculated. A normal score transform of the categorical sample data is then performed, honouring each category's prior probability. The values calculated from the normal score transform are then assigned to each sample data point. The variogram produced for estimating the degree of dissimilarity between points thus refers to these normal score values of the categorical data. Steps 1–5 from above then proceed, of which the randomly drawn value is thus also a normal score value. Back transformation of the normal score value is then constrained to respect the relevant discrete set of the categories in the original data (e.g. [1, 2, 3]).

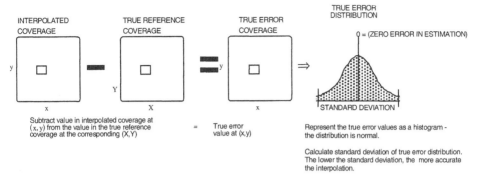

Figure 1.4 'True error' found by subtracting the interpolated estimate from the reference coverage, which is assumed to be error-free.

1.4.5 Measuring interpolation accuracy

From the number of equally 'good' simulations of the full spatial extent of an attribute under study, a single map may be selected either at random from the set of realizations, in accordance with some defined 'goodness' criteria, or be an average composite of all realizations. This project created ten simulations for each experiment, randomly selecting a representative sample of four of these to compare interpolation accuracy against the single 'best' Thiessen polygons and kriging realizations. A composite of the ten simulations was also created and compared to the results from the other interpolation methods in each experiment. For continuous and spatially variable data sets the composites represented the mean value of all simulated values for each location. For categorical data sets the composite image represented the modal value of all simulations for each location.

The accuracy of each interpolation method for the continuous and spatially variable reference data sets, different sample sizes and sample distributions was compared by calculating the standard deviations of each 'true error' distribution (see Figure 1.4). True error refers to the difference between the values interpolated at locations and the exact attribute value at the corresponding location in the real world. In calculating each true error distribution the original sample point values were not included. This corrected for any bias that might result from different sample sizes. The lower the standard deviation of the true error distribution, the closer the interpolated values were to the 'true' pattern, and the more accurate the interpolation was judged to be.

For categorical data a calculation of the standard deviation cannot be applied. Instead, the percentage of pixels that had been classified correctly was calculated and used to compare between tests.

1.4.6 Measuring the estimation uncertainty in the interpolations

The second part of our study sought to develop a measure of the uncertainty of the estimations produced by interpolation. Ideally, we would like this measure of uncertainty to be mappable. Then, if we can show that areas of high estimation uncertainty correspond closely with areas where we know the true error to be high, we will have a technique which allows us to interpret and quantify the spatial variations in the uncertainty throughout the interpolated map.

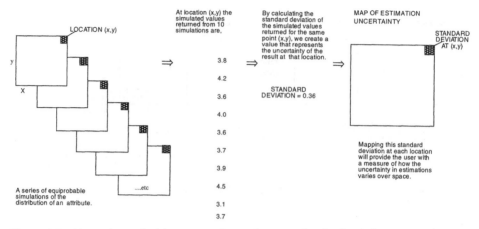

Figure 1.5   Measuring estimation uncertainty using a stochastic simulation approach.

The kriging variance

Kriging variance is calculated by comparing the variance of the sampled values within the search neighbourhood with the average variance between the location of the interpolated point and all sample points in that neighbourhood (see Isaaks and Srivastava, 1989, and Bailey and Gatrell, 1995, for examples on calculating kriging variance). With each ordinary kriging estimate using GSLIB, the kriging variance for each pixel is calculated simultaneously.

Modelling estimation uncertainty using stochastic simulations

Stochastic simulation algorithms create a user-specified number of equiprobable realizations of the attribute under study. A reasonable means to measure the uncertainty in estimations is to compare the values at each location in each of these different simulations. At those locations where the resultant value is similar, or the class code is the same over many realizations, the more certainty we may attach to the result.

For continuous and spatially variable data sets it is proposed that the standard deviation of the simulated values returned for each individual pixel over a series of realizations provides a method of measuring uncertainty which is more reflective of the true error than the kriging variance. This means that the higher the standard deviation at an unsampled location, the more uncertain we are of the estimate produced here (see Figure 1.5). A map showing this standard deviation at every location based on the simulation results would also be quite simple to map and comprehend. This map can then be compared with one showing the kriging variance at each location to determine which is a closer and more realistic representation of the map of true error.

For categorical data sets this process of measuring uncertainty is not appropriate. The value returned at any one location accords with a code that represents a category of data. Any calculation of standard deviation based on these category codes would be meaningless as a measure of uncertainty. Instead maps of uncertainty were created by:

- Randomly selecting a single 'good' realization or composite realization of the data.

- Comparing the classification of each pixel for every simulation against this chosen realization. This would return a value of 0 (the pixel value from the chosen realization matches the classification for the same point in an equiprobable simulation) or 1 (the pixel classification did not match). Totalling these 0's and 1's for all compared simulations (n), and dividing by n would return a value describing the uncertainty of each pixel value for the chosen realization. The lower the value, the more often the pixel is assigned the same class in different realizations, and the higher the confidence we have in the simulated value.

Using this technique a map showing uncertainty was produced for the categorical simulations, which could then be compared against a map showing true error.

1.5 RESULTS

1.5.1 Measuring spatial variability and variogram modelling

Experimental semivariograms for small sample sizes (i.e. 25 and 49 data points) in many cases were not stable enough to be confidently modelled. Consequently, ordinary kriging and stochastic simulation realizations of these smaller data samples could not be produced.

For all three types of geographical reference data, most variograms that could be modelled used either a single spherical or single exponential structure. In experiments where a nested structure was used, a combination of spherical modelling and exponential modelling was performed. In no experiments was a Gaussian structure employed.

1.5.2 The relative accuracy of stochastic simulation, ordinary kriging and Thiessen polygons for different sample sizes and distributions of geographical data

For all types of reference data, a single randomly selected simulation was generally the least accurate method of interpolation. However, by creating mean or modal composites from ten individual simulations, the accuracy of the interpolation was consistently improved. For continuous and spatially variable reference data sets (see Figures 1.6(a) and 1.6(c) respectively), the accuracy of the mean composite simulations was generally better than Thiessen polygons, but less accurate than interpolation by ordinary kriging. In one case, a mean composite simulation using 289 randomly distributed points from the spatially variable reference data set produced a mean composite that was marginally more accurate than ordinary kriging. For the spatially variable data set it was noted that the accuracy of each interpolation method did not always improve with increasing sample size.

For categorical data Thiessen polygons produced the most accurate interpolations. Interpolation by Thiessen polygons generally classified 7%–16% more pixels correctly than any individual stochastic (PDF) simulation. However, when modal composites of the simulations were created, the accuracy markedly improved and approached that of the Thiessen Polygons (see Figure 1.6(b)).

For continuous and categorical sample data, systematic distributions of data tended to produce realizations of the highest accuracy for all interpolation methods (see Figure 1.6). This might be expected from a knowledge of sampling theory in that a systematic sample should better detect a continuous trend or regional variations, whereas a random

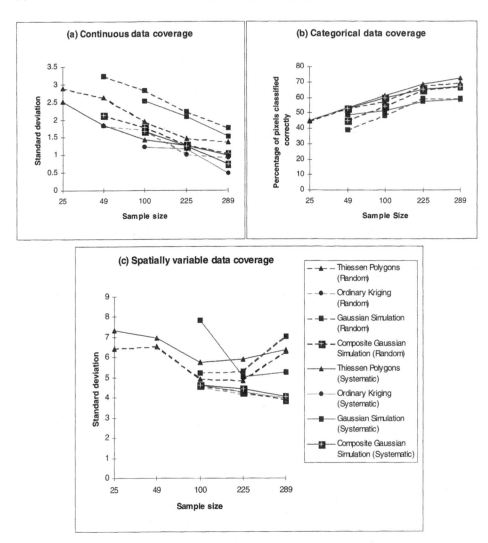

Figure 1.6 Interpolation accuracy for Thiessen polygons, ordinary kriging (for non-categorical data sets), randomly selected simulations, and composite simulations of (a) a continuous data coverage, (b) a categorical data coverage, and (c) a spatially variable data coverage. Graphs (a) and (c) show a comparison of the standard deviation of the distribution of true error (estimate minus true value for 2 500 pixels), and graph (b) shows a comparison of the percentage of pixels that were classified correctly. All graphs are plotted for the range of sample sizes and show the differences between using a random or systematic distribution of data points. Note that the scaling on the x-axis is non-linear to allow a compact diagram.

sampling method might undersample in certain areas. For the spatially variable data sets, interpolations were marginally more accurate when the sample data were randomly distributed.

1.5.3 The accuracy of the interpolation techniques for attributes of different geographical nature

In this section we consider the results of the tests in more detail. Based on the results of Figure 1.6 we focus discussion on the case of the larger sample size (289 points). We consider the systematically distributed data for the continuous data sets and the random distribution of points for the spatially variable and categorical data sets. These results are selected from Figure 1.6 as they represent, in each case, the density arrangement of data that generally provided the best accuracy. In this section we also seek to understand the relative differences in accuracy due to the methods themselves – when (for all interpolation techniques) conditions of density and geometrical arrangement of data were held constant.

Turning first to the continuous reference data, Figure 1.7 shows realizations produced by Thiessen polygons, ordinary kriging, and Gaussian simulation using 289 systematically distributed points of the continuous reference data. The Gaussian simulations shown are two randomly selected realizations from the 10 produced, and the mean composite of these ten simulations. The Thiessen polygons coverage displays the oversimplification of reality that the method often creates: the polygon boundaries show a sudden change in the attribute values over space rather than representing the true spatial nature of the phenomenon (which in the continuously distributed reference data set is that of gradual change). The systematic distribution of the sample points in this example also serves to illustrate how the sample point locations strongly determine the size and shape of each Thiessen polygon.

The ordinary kriging realization for 289 systematically distributed points displays the smoothing effect associated with this interpolation method. However, for the phenomena that are to be estimated this smoothing characteristic appears advantageous – both visually capturing the trend and accurately estimating the values of the attributes. In comparison, randomly selected Gaussian simulations appear 'noisy'. Yet the spatial distribution of estimated values generally matches the patterns shown in the continuous attribute reference map. The composite simulation, as calculated and displayed in Figure 1.6(a), appears more accurate than either of the randomly selected simulations.

Realizations were then produced from a sample of 289 randomly distributed points of categorical land cover data using Thiessen polygons and categorical (PDF) simulation. The general characteristic of Thiessen polygons to create a specific attribute's 'area of interest' around a sample location tended to respect the spatial distribution of the land cover reference data when the sample size was large. This is supported in Figure 1.6(b) where interpolation by Thiessen polygons was shown to be more accurate than categorical simulation in all cases. The randomly selected categorical (PDF) simulations appeared 'noisy', although the broad pattern of the spatial distribution of the land cover categories was generally displayed in each simulation. The effect of combining simulations seemed to improve the interpolation accuracy by reducing a large amount of the noise. It was hypothesized that using a mode composite based on 30 simulations of this categorical data might produce an interpolation that was more accurate than interpolation by Thiessen polygons. However, constructing a mode composite of 30 categorical simulations using

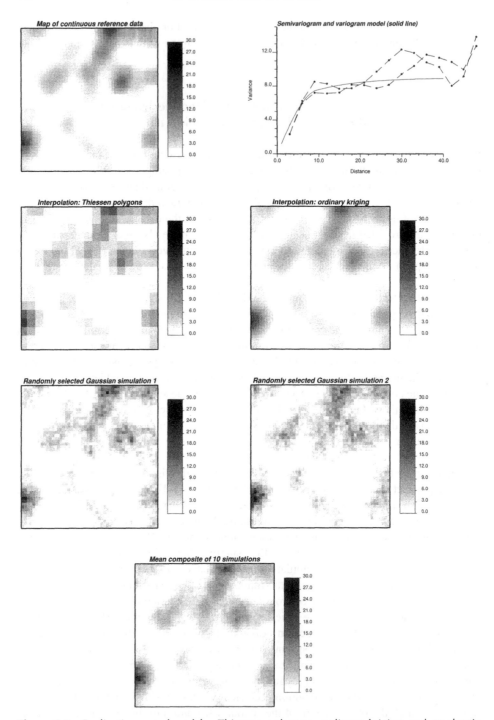

Figure 1.7 Realizations produced by Thiessen polygons, ordinary kriging and stochastic simulation from a systematic sampling of 289 observations of a continuous property (e.g. elevation).

the sample of 289 randomly distributed data points only improved the percentage of pixels classified correctly by 1.4%. This suggests that, while combining categorical simulations does improve interpolation accuracy by reducing the noise evident in individual simulations, the method's focus on attempting to recreate local variability does not relate well with those geographical data sets where the distribution of information is into discrete areas of uniform attribute value.

For the case of the spatially variable reference data, a comparison was made of interpolations by Thiessen polygons, ordinary kriging, and a stochastic Gaussian simulation for 289 randomly distributed data points. Thiessen polygons, by their nature, create an area of interest from a sample of one. This characteristic highlighted the inability of this interpolation method to capture the local variability of this reference data set. Ordinary kriging, even for a data set of 289 points, produced a smoothed surface. While the general trend in the attribute values was captured, the local variability in attribute values was not reproduced. The randomly selected Gaussian simulations appeared visually to give a better reproduction of the reference data (i.e. local variability in the attribute appeared to have been captured). Using the per-pixel test of accuracy against the 'true' reference data, however, these individual Gaussian simulations were less accurate than either of the realizations produced from ordinary kriging or Thiessen polygons. Yet, the mean composite of the Gaussian simulations from a random sample of 289 data points, as shown in Figure 1.6(c), was the most accurate of all the realizations. Our interpretation of these results is that the effect of creating a mean composite from the individual stochastic simulations was to preserve local variability, while at the same time the averaging of values over ten simulations reduced the noise and hence improved the overall accuracy.

1.5.4 The ability of interpolation methods to reflect estimation uncertainty

Figure 1.8 shows the kriging variance created for the interpolations of 49 randomly distributed points and for 289 systematically distributed points of the continuous reference data. These maps should be compared with the true error calculated for each respective interpolation. This figure illustrates the dependence of the kriging variance upon the geometrical arrangement of the sample points. In relation to the true error, the kriging variance is a poor measure of uncertainty, stating nothing more than what common sense would suggest – the further one moves away from a data point or set of data points, the less accurate the estimate will be. This is especially highlighted in the kriging variance map of 289 systematically distributed sample data. The pattern of kriging variance is also independent of the magnitude of the values at the observation points. This is shown by the grey scales required to display each kriging variance map adequately. For example, the map of kriging variance of 49 randomly selected points would appear completely black (except for the white dots showing the location of the original sample points) if the grey scale was aligned to that of the true error maps. Hence, in terms of both spatial patterns and the magnitude of the measure produced, kriging variance is a poor substitute for a map of true error.

Models of uncertainty were also created from the ten equiprobable simulations of the equivalent 49 and 289 point sets of continuous data (see Figure 1.8). These simulated maps of uncertainty reflect visually the true error maps produced for each composite image. Furthermore, the grey scale values used to display each estimate of uncertainty compare directly to the grey scale values of the true error maps. For the spatially variable data a similar correspondence was found between the measure for estimation uncertainty

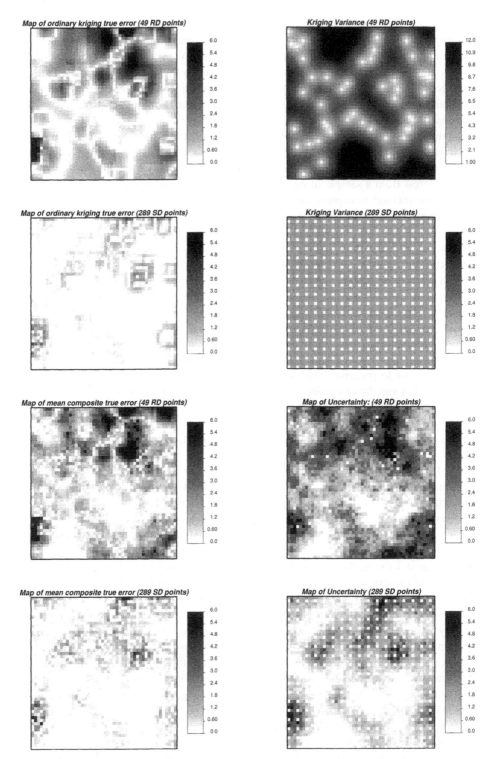

Figure 1.8 Measuring uncertainty in the estimations: spatially continuous data sets. The figure shows the kriging variance and maps of uncertainty produced using a simulation method, alongside their respective true error maps. The simulation approach consistently shows a markedly closer resemblance to true error for different sampling densities and the different geometrical arrangements of sample points.

derived from the simulation composites and the map of true error. As with the maps showing estimation uncertainty for the continuous data sets, the analyst could identify the areas where estimations are most uncertain, and using the scale give some approximate numerical value as to the likely level of inaccuracy.

We may recall that a different method was required to derive a measure of estimation uncertainty for comparison with a true error map for the categorical data. The approach adopted to model uncertainty for categorical data was to return a probability (based on all simulated values for one location) of the simulated pixel being incorrect. While this measure of estimator uncertainty did not match the true error map exactly, the results obtained suggest the approach can certainly be used to identify areas of relative confidence and uncertainty in a stochastically simulated interpolation. Areas where the measure of uncertainty was low generally corresponded with areas in the true error map where the pixels were correctly classified. Inclusion of confidence thresholds set at certain probabilities that reflect the uncertainty in the estimates may produce an uncertainty map that more closely matches a map of true error.

While we have reported here on the results of measuring estimation uncertainty using a simulation approach for only four of the data sets, these results were consistent for all other sample sizes and sample distributions. The method is suitable also for individual simulations when other equiprobable realizations are used to create the maps of uncertainty – see Chainey (1996). Hence, in all cases, an approach based on simulation provided a better map of estimation uncertainty than an approach based on kriging variance.

1.6 DISCUSSION

There is an increasing desire in the GIS community to improve the accuracy of interpolation, and also a wish to indicate the uncertainty inherent in these estimated data sets. For continuous and spatially variable data kriging methods are widely used and are considered to be one of the more accurate methods for producing an interpolation showing the full spatial extent of an attribute. The relatively high interpolation accuracy of kriging is confirmed in this study, particularly where the attribute to be interpolated is generally continuous in its spatial variation (e.g. elevation). Realizations produced by stochastic simulation were generally less accurate than kriging, but in most cases were more accurate than using Thiessen polygons. The one experiment where a stochastic simulation produced a realization that was marginally more accurate than by kriging was on a geographical reference data set that was spatially variable. Deutsch and Journel (1992) note that a stochastic simulation approach tends to respect the local statistical variability and the global texture of the data. It appears from these experiments that, when a spatially variable data set is to be interpolated, the ability of a stochastic simulation method to honour local variability relates well to the real world distribution of this type of geographical information.

This idea of performing the appropriate interpolation technique based on the underlying spatial variation of the attribute to be interpolated is further emphasized by the realizations produced for categorical data. The nature of the interpolation by Thiessen polygons – to create discretely bounded areas of uniform attribute values – appears advantageous if the geographic phenomena itself occurs as uniform patches and is not locally variable.

Creating a single composite map based on a series of simulations has been likened to recreating the results from a kriging algorithm (Deutsch and Journel, 1992). The argument is that when a large number of simulations of an attribute are used, the average value at any one location should closely approximate to the single estimated value returned by kriging (i.e. the effect of averaging the simulated values for each location is to create a smoothed map which is usually similar to the kriged result). However, if the user has control over the number of simulations used to create a composite, the use of a composite simulation has significant additional advantages for data sets commonly used by the GIS community. Modal composites of categorical data were shown to reduce the noise of the individual simulations and to produce an interpolated coverage of a definite higher accuracy. Using a small number of simulations (i.e. 10) for spatially variable data, the effect appeared to create a coverage that closely resembled the local variability of the attribute, smoothed any large outliers in the simulation results, but did not 'oversmooth' the data in the way the kriged map did. This may explain why the mean composite of 289 spatially variable, randomly sampled points was a more accurate realization than either Thiessen polygons or kriging.

The experiments conducted have also shown the flexibility of a stochastic simulation approach in the way realizations could be created for all three kinds of reference data sets. Although any individual simulation was generally the least accurate, the prospect of improvements from other stochastic simulation algorithms (e.g. a Boolean algorithm for categorical data that simulates the geometric shape of an attribute) suggests further potential in the technique. Further improvements in the accuracy of stochastic simulation realizations beyond what was achieved in this study may result from:

■ The adoption of a hybrid approach that uses a combination of different stochastic simulation algorithms to recreate more effectively the wide variety of features and statistics found in some geographical attributes (Deutsch and Journel, 1992).

■ Modification of the sampling distribution. Laslett (1994) noted that kriging on randomly distributed samples and then adding further points in areas of high attribute value produces considerably more accurate realizations. A similar approach can be adopted using stochastic simulation, although care must be taken as this clustering approach may significantly affect the global mean and variance (Deutsch and Journel, 1992).

■ More detailed modelling of the experimental variogram. This is particularly true for categorical data where each category could be modelled individually for its spatial variability.

The empirical studies in this project have confirmed an emerging opinion that the kriging variance is ineffective as a method for describing the uncertainty in the estimations. In all cases a map of kriging variance provided neither a qualitative picture nor a useful quantitative estimate of interpolator uncertainty and never closely resembled a map of 'true error'. Given that the true error map is not normally available, the results from this study suggest that a series of stochastic simlations may provide a more easily understood and computable measure of the uncertainty in an interpolation estimate.

While stochastic simulation can be used as an interpolator in its own right, the study has, however, shown that stochastic simulations do not usually produce the most accurate interpolation result for all types of data. Do we then have to sacrifice interpolation accuracy for a better measure of estimation uncertainty? Kriging and stochastic simulation come from the same family of interpolation techniques – geostatistical interpolation. Each method requires the same inputs – the same sample data and the same variogram

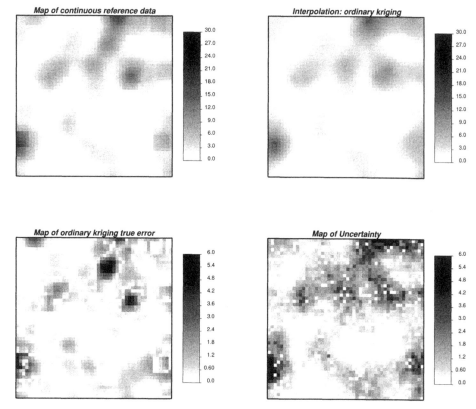

Figure 1.9 Kriging consistently produces an accurate interpolated coverage, while a simulation method of showing uncertainty is a reasonable surrogate of true error (which is not normally available). As kriging and stochastic simulation are based on the same principles and require the same inputs, it is suggested that the two methods could be combined to provide the GIS analyst with a better quality result.

model. It is thus possible to suggest that a combination of the two geostatistical interpolation methods – kriging to accurately interpolate a set of values, and stochastic simulation to supply the map of uncertainty in the estimations – may be a viable means to assist the GIS user with an improved assessment of the quality of their interpolation result. This complementary approach is illustrated in Figure 1.9. As a realization representing the full spatial extent of the attribute from a sample of 289 randomly distributed data points, the kriged map is the most accurate example produced using the interpolation methods in this study. Using the same sample data and the same variogram model, a stochastic simulation can show the range in possible values at each unsampled location. The larger the range, the more uncertain we are in any estimation produced at each unsampled location generated from a geostatistical method. Showing this as a map will, therefore, assist the GIS user to make a value judgment on the reliability of the kriged estimates and on any subsequent analyses performed on the interpolated coverage.

Although methods based on stochastic simulation would seem to offer considerable opportunities for interpolation and for improved visualization of error, the limitations of the technique need to be considered. For example, the difficulties in modelling the experimental variograms for small sample sizes highlights a general weakness of geostatistical interpolation methods. In contrast, realizations were produced using Thiessen polygons for all sample sizes, although whether the Thiessen polygons from small sample sizes supply the necessary accuracy required for subsequent GIS operations is open to question. The results from this project suggest that the expense of sampling a larger number of locations in order to allow the adoption of geostatistical interpolation methods can usually be justified by the improved accuracy and a measure of uncertainty in the result.

1.7 CONCLUSION

Measures of the confidence that can be placed in spatial data sets are becoming increasingly imperative in the GIS community. While there is greater awareness of the 'false lure of the graphic image', the practical need to transform sample points to surfaces means that interpolated results are still often accepted without query and may then go on to be assumed to be correct for subsequent GIS operations. This study aimed to investigate one interpolation technique that is receiving increasing attention by the GIS community: stochastic simulation. The potential of the method was evaluated by comparison with two of the more commonly used GIS interpolation techniques – Thiessen polygons and kriging.

An experimental approach was followed which allowed the relative performances of the three interpolators to be compared under different conditions of data density and geometry. The results were consistent with expectations from a knowledge of sampling theory in that the accuracy of interpolation by stochastic simulation generally improved with sample size; systematic distributions of sample points usually captured more accurately the trend in continuous data sets; and random distributions produced slightly more accurate results when the sample data were spatially discontinuous. Calculations of interpolation accuracy showed that, for a categorical land cover reference data set, individual stochastic simulations were not particularly accurate, but a mean composite produced from a series of simulations improved the interpolation accuracy compared to that obtained by Thiessen polygons. For the continuous and spatially variable reference data sets stochastic simulations were generally less accurate than kriging, but more accurate than Thiessen polygons.

These results support the recent adoption of kriging as an interpolation technique into several GIS and associated spatial analysis packages. An additional attraction of kriging has been the claim that the kriging variance can supply a measure of the uncertainty in the resulting map. However, this paper has shown that kriging variance is relatively ineffective as a surrogate for the true error in interpolated estimates. The use of a series of stochastic simulations did, however, illustrate more accurately estimation uncertainty, producing maps that closely resembled both the spatial patterns and magnitude of values on the calculated maps of true error.

Taken as a whole, the results of this experimental study suggest that stochastic simulation provides flexibility in the type of geographical data sets which can be interpolated. We have illustrated its use for attributes measured on both categorical and interval scales and for data sets with differing degrees of spatial variability. While the basic stochastic simulation methods used here are not consistently as accurate as kriging for interpolation, the study exposes the potential of the technique as a spatial interpolator. More important, however, stochastic simulation is shown to offer an accurate source of information about the uncertainty of the estimations it has produced. Being based on the same geostatistical principles as kriging, stochastic simulation should perhaps be considered not just as an alternative method of interpolation, but more powerfully as a complimentary method which provides information about the uncertainty in the estimates produced by geostatistical methods.

References

ATKINSON, P. and KELLY, R. (1996) Mapping with uncertainty, *Mapping Awareness*, **10** (8), 35–37.

BAILEY, T.C. and GATRELL, A.C. (1995) *Interactive Spatial Data Analysis*, Harlow: Longman Scientific and Technical.

BIERKINS, M. and BURROUGH, P.A. (1993) Stochastic indicator simulation, Parts I and II. *J. Soil Science*, **40**, 477–492.

BURROUGH, P.A. (1986) *Principles of geographical information systems for land resource assessment*, Oxford: Oxford University Press.

BURROUGH, P.A., VAN RIJN, R. and RIKKEN, M. (1996) Spatial data quality and error analysis issues, in: M.F. Goodchild, L.T. Steyaert, B.O. Parks (Eds), *GIS and Environmental Modelling: Progress and Research Issues*, Fort Collins: GIS World Books.

BUTTENFELD, B.P. (1993) Representing data quality, *Cartographica* – special issue on mapping data quality **30** (2&3), 1–7.

CARBONE, G.J., NARUMALANI, S. and KING, M. (1996) Application of remote sensing and GIS technologies with physiological crop models, *Photogrammetric Engineering and Remote Sensing*, **62** (2), 171–179.

CHAINEY, S.P. (1996) Stochastic simulation: an alternative interpolation technique for digital geographic information, *Dissertation for MSc* in *Geographical Information Systems*, Department of Geography, The University of Edinburgh, Scotland.

DEUTSCH, C.V. and JOURNEL, A.G. (1992) *GSLIB: Geostatistical Software Library and User's Guide*, Oxford: Oxford University Press.

DUBRULE, O. (1983) Two methods for different objectives: splines and kriging. *J. International Association of Mathematical Geology*, **15**, 245–255.

DUBRULE, O. (1984) Comparing splines and kriging, *Computers and Geosciences*, **101**, 327–338.

DUBRULE, O. (1989) A review of stochastic models for petroleum reservoirs, in Armstrong (Ed.), *Geostatistics*, The Netherlands: Kluwer Academic.

ENGLUND, E.J. (1993) Spatial Simulation: Environmental Applications, in M. Goodchild, B. Parks, and L. Steyaert (Eds), *Environmental Modelling with GIS*, Oxford: Oxford University Press.

FISHER, P. (1994) Hearing the reliability in classified remotely sensed images, *Cartography and GIS*, **21** (1), 31–36.

GOODCHILD, M., PARKS, B. and STEYAERT, L. (Eds.) (1993) *Environmental Modelling with GIS*, Oxford: Oxford University Press.

GOTWAY, C.A. and RUTHERFORD, B.M. (1996) The components of geostatistical simulation, *2nd Int. Symp. on Spatial Accuracy Assessment in Natural Resources and Environmental Sciences*, 30–38, Fort Collins, CO.

HALDORSEN, H. and DAMSLETH, E. (1990) Stochastic modelling, *J. Petroleum Technology*, April, 404–412.

ISAAKS, E.H. and SRIVASTAVA, R.M. (1989) *Applied Geostatistics*, Oxford: Oxford University Press.

JOURNEL, A.G. (1986) Geostatistics: Models and tools for earth sciences, *Journal of the International Association for Mathematical Geology*, **18** (1), 119–140.

JOURNEL, A.G. (1996) Modelling uncertainty and spatial dependence: stochastic imaging, *Int. J. Geographical Information Systems*, **10** (5), 517–522.

JOURNEL, A.G. and ALABERT, F. (1990) New method for reservoir mapping, *J. Petroleum Technology*, February, 212–218.

LAM, N.S. (1983) Spatial interpolation methods: a review, *American Cartographer*, **10**, 129.

LASLETT, G.M. (1994) Kriging and splines: an empirical comparison of their predictive performance in some applications, *J. American Statistical Association*, **89** (426), 391–409.

LASLETT, G.M. and MCBRATNEY, A.B. (1990) Further comparison of spatial methods for predicting soil pH, *J. Soil Science Society of America*, **54**, 1553–1558.

MACEACHREN, A.M., HOWARD, D., VON WYSS, M., ASKOV, D. and TAORMINO, T. (1993) Visualizing the health of Chesapeake Bay; an uncertain endeavour, *Proceedings, GIS/LIS '93*, Minneapolis, 2–4 Nov. 1993, pp. 449–458.

MYERS, D.E. (1996) Choosing and using simulation algorithms, *Second International Symposium on Spatial Accuracy Assessment in Natural Resources and Environmental Sciences*, Fort Collins, CO, 23–29.

OLIVER, M.A. and WEBSTER, R. (1990) Kriging: a method of interpolation for geographical information systems, *Int. J. Geographical Information Systems*, **4** (3), 313–332.

ROSSI, R.E., BORTH, P.W. and TOLEFSON, J.J. (1993) Stochastic simulation for characterizing ecological spatial patterns and appraising risk, *Ecological Applications*, **3**, 719–735.

SRIVASTAVA, R.M. (1996) An overview of stochastic spatial simulation, *Second International Symposium on Spatial Accuracy Assessment in Natural Resources and Environmental Sciences*, Fort Collins, CO, 13–22.

STOYAN, D., KENDALL, W. and MECKE, J. (1987) *Stochastic Geometry and its Applications*, New York: John Wiley and Sons.

VAN DER WEL, F., HOOTSMANS, R.M. and ORMELING, F.J. (1994) Visualization of data quality, in A.M. MacEachren, and D.R.F. Taylor (Eds), *Visualization in modern cartography*, London: Pergamon, 313–333.

VAREKAMP, C., SKIDMORE, A.K. and BURROUGH, P.A. (1996) Using public domain geostatistical and GIS software for spatial interpolation, *Photogrammetric Engineering and Remote Sensing*, **62** (7), 845–854.

VOLTZ, M. and WEBSTER, R. (1990) A comparison of kriging, cubic splines, and classification for predicting soil properties from sample information, *J. Soil Science*, **41**, 473–490.

WARD, R.C. and ROBINSON, M. (1990) *Principles of Hydrology*, London: McGraw Hill.

WEBSTER, R. and OLIVER, M.A. (1990) *Statistical Methods in Soil Science and Land Resource Survey*, Oxford: Oxford University Press.

ZHANG, J. (1996) A surface-based approach to the handling of uncertainties in an urban-oriented spatial database, Unpublished Ph.D. thesis, Department of Geography, The University of Edinburgh, UK.

CHAPTER TWO

Spatial evolutionary algorithm

ROMAN M. KRZANOWSKI AND JONATHAN F. RAPER

2.1 INTRODUCTION

Evolutionary modelling methods have gained broad recognition by many researchers engaged in divergent scientific fields (Davis, 1991; Goldberg, 1989; Holland, 1993). Indeed, several hundreds of applications of evolutionary based techniques bear witness to the cogency of this modelling approach (Liepins and Hillard, 1989; Michalewicz, 1992; Mitchell, 1996; Nissen, 1993).

Evolutionary models of spatial problems have been investigated since the late 1980s and several successful applications have been developed during that time. Delahaye *et al.* (1994) implemented a real-value coded evolutionary algorithm for the optimal design of flight corridors for aircraft flight-management. Hosage and Goodchild (1986) and Bianchi and Church (1992) investigated p-median models using canonical and integer coded genetic algorithms. Openshaw (1988 and 1992) investigated spatial interaction models and data mining techniques using evolutionary methods. More recently, the same author, together with Turton (1996) investigated applications of genetic programming in spatial modelling. Pereira *et al.* (1996) presented a very complete spatial suitability model based on canonical genetic algorithms. Hobbs (1993, 1994 and 1995) developed several innovative applications of evolutionary methods for spatial aggregation of polygons and networks. Brooks (1996) proposed the application of a hybrid genetic algorithm for spatial modelling. Dibble (1994) proposed the use of genetic methods for knowledge discovery in spatial data, and Dibble and Densham (1994) explored the use of evolutionary paradigm for the multiobjective modelling in SDSS. Keller (1995a and 1995b) developed a line generalization approach based on genetic model. And Cooley and Hobbs (1992) investigated genetic algorithms for the classification of continuous data.

The research on spatial evolutionary models, spanning almost a decade, has demonstrated that those modelling techniques can be effectively applied to a variety of spatial problems. What is needed now – more than new applications – is an evolutionary modelling framework that could be reused across a spectrum of spatial models. Thus, rather than concentrating on one problem, in this paper we formulate a general framework for spatial modelling (or, at least, for the modelling of the broad class of spatial systems) with an evolutionary paradigm. The proposed modelling framework is called a spatial evolutionary algorithm (SEA). The following paragraphs present the elements of SEA, describe the development of the spatial models using SEA and, thereafter, set out an evolutionary model of a planar coverage problem (PCP) developed with SEA.

2.2 SPATIAL EVOLUTIONARY ALGORITHM – COMPONENTS

2.2.1 Representation of spatial systems and objects

SEA is designed to model spatial systems – that is, systems of objects distributed in space and defined over some spatial information. Spatial systems are composed of objects, where each object has a location in space and other properties like an area of influence or size. In addition, spatial systems and their component objects have associated measures of quality that permit one to distinguish a better system (or its component objects) from an inferior one with respect to some objectives.

The basic component of SEA – *organism* – represents an object in a spatial system. Each object-organism has a set of attributes describing its properties. Attributes of the object-organism are coded in the object-organism genetic material; one attribute is coded in one gene. Organisms form a *population of organisms*. A *population of organisms* represents a *system of objects*. A set of populations of organisms represents a *hyper-population*. This hierarchy of SEA components is given in Figure 2.1. The proposed representation of the spatial model as a hierarchy of objects resembles the object-oriented view of the environmental processes developed by Raper and Livingston (1996).

2.2.2 Representation of spatial information

Spatial information in SEA is represented in a layered model with each layer representing a particular aspect of the space. Some other possibilities of representation of spatial information that can be used in evolutionary models are discussed by Raper (1989).

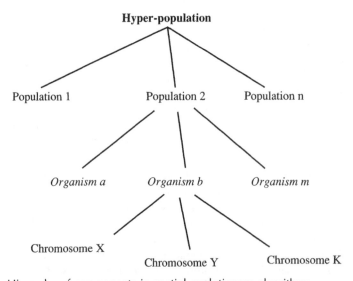

Figure 2.1 Hierarchy of components in spatial evolutionary algorithms.

2.2.3 Spatial evolutionary operators

The SEA has been designed with eight operators: initialization, selection, fitness, mating, cross-over, mutation, learning, and objective function. Because of the hierarchical architecture of SEA (organisms–populations–hyper populations: see Figure 2.1) each operator with the exception of the objective function, has two levels of implementation: population and organism level. Each operator may also be implicit and explicit. An *implicit operator* is an operator that is not implemented *per se*, but whose action is an observable result of the activation of other operators. Although differentiating operators as being *implicit* or *explicit* may seem to add an undesirable level of complexity, the benefit is seen in the consistency of the definition and in the description of the SEA framework, making it more complete and more flexible. SEA operators are, conceptually, extensions of traditional evolutionary operators. However, the structure of SEA makes its operators quite different from its non-spatial counterparts. Also, because spatial evolutionary models have chromosomes representing different features, all the operators affecting chromosomes are chromosome-specific.

The SEA operators are defined as follows:

- An *initialization operator* is an operator that generates the initial hyper-population. The hyper-population is created by generating a predefined number of populations. A population is created by generating a predefined number of organisms. An organism is generated by pooling a uniform random number from a (0,1) range for each feature of the organism, and then mapping it into the feature domain, and coding into the chromosome.

- A *selection operator* on the level of populations is implemented as a traditional selection process defined in GA literature. On the level of organisms a selection is implicit. That is, selected organisms are those that belong to selected populations.

- *Fitness operators* in SEA are implemented as focal (on an organism level) and zonal (on a population level) operators of map algebra (Tomlin, 1985). Map algebra operators have originally been defined on the operational level, and as such they are difficult to incorporate into the abstract framework (such as SEA). However, they do provide a convenient metaphor for spatial operators, a metaphor that is easily understood and operationalized.

- A *mating operator* on the level of populations is a process in which two populations are selected at random from a pool of selected populations. On the level of organisms a mating operator implements the process in which two organisms from two different populations, selected by the population-mating operator, are selected on the basis of some measure of proximity.

- A *cross-over operator* on the level of populations is implicit: it occurs when objects in mating populations exchange genetic material. The cross-over of populations takes, as input, two populations and generates one offspring population. At the level of organisms the cross-over takes as an input two organisms selected in mating and produces one organism. A cross-over is chromosome-specific.

- SEA has two types of *mutation* (for organisms and for populations): 'Big M' mutation and 'small m' mutation. 'Big M' mutation affects the whole population. It is implemented as a random change for all the genetic material of all of the organisms in a randomly selected population. 'Small m' affects a single organism. It is a random change of the genetic material of the randomly selected organism from a randomly selected population.

- *Learning* in SEA is an operator that improves the fitness of organisms in a population (and the fitness of a population). The actual implementation of the learning operator is problem-specific.

- *Objective function* in SEA is defined only for populations (as only the population fitness maps into the problem space) and is equivalent to the population's fitness. It expresses the fitness of the evolutionary units with respect to the problem objectives.

2.3 SPATIAL EVOLUTIONARY ALGORITHM – IMPLEMENTATION

Implementation aspects of SEA are discussed in this paper within the context of the Planar Covering Problem. It has to be stressed that SEA is a framework for the design of evolutionary models of spatial problems and, as such, can accommodate many different models on many levels of complexity. Planar Covering Problem is used here only to provide an example of how to operationalize the SEA framework components and it should not prevent anyone from understanding SEA's capabilities for modelling more complex PCP or different spatial problems altogether. In this study the PCP problem is defined as follows:

given a set of n facilities, each with a circular service area of a constant radius r, define the locations of those facilities on a demand surface s, such that they service maximum demand possible, with T(s) =const

The elements of a model of the planar covering problem include: individual facilities, sets of facilities, properties of facilities, problem constraints, and the spatial environment over which the PCP models are defined. In a spatial evolutionary framework, individual facilities are represented as organisms. Properties of facilities (location, size, serving area) are the organisms' features and they are coded into its chromosomes. A set of facilities is regarded as a population, the problem constraints are expressed via fitness function, and a spatial environment of the PCP is represented as a set of the spatial data layer.

The genetic makeup of organisms represents the position (X and Y coordinates) of the organisms in a metric space (no other features of organisms are considered in this version of PCP). Each coordinate is defined by a separate chromosome. Chromosomes have natural coding. Natural coding, in contrast to binary coding, uses – as chromosome values – the actual values of the represented attribute (location, size, etc.).

Fitness for organisms and populations is defined by focal and zonal operators of the map algebra. Focal operator for the planar covering problem is defined as a sum of attributes* assigned to points that belong to the focal neighborhood of the organism; in the case of PCP the focal neighbourhood is a circle of a radius r. Zonal operator for the planar covering problem is defined as the sum of attributes assigned to points belonging to a zone <l>. Zone <l> is defined as a sum of subspaces assigned to the objects of a population. A selection operator for population is implemented as Tournament 1–1 (Michalewicz, 1992). Selection for organisms is implicit – i.e. organisms selected are those that are in selected populations. The *mating* operator for organisms can be described as follows: for every organism i in a population 1 (a_{i1}) an organism in a population 2 (a_{k2}) that is closest to it by some measure D is selected. As a measure of closeness

---

* Attributes are expressed as spatial data layers.

the implementation of SEA for PCP problem employed Euclidean distance metric. A *cross-over* operator for organisms has two implementations – fixed and random point–both defined by the formula:

$$l_{ik} = w^1 l_{1k} + w^2 l_{2k}$$

When weights $w^1 = w^2$ are equal 0.5, this formula represents a fixed point cross-over. A random point cross-over has weights (w) implemented as uniform random numbers from a <0,1> range – rnd(0) and 1-rnd(0). The l_{ik} are values of genes entering the cross-over operator with k being a gene responsible for a particular feature and i representing an organism gene belongs to. Mutation for location chromosomes is defined over a metric space and depends on the current state of the chromosome. Mutation is controlled by the probability of mutation (separate for big and small mutation) which expresses the probability with which the mutation operator is activated during the evolution. The *learning* operator is based on a local, deterministic hill-climbing and is designed after the Tornqvist algorithm (Tornqvist *et al.*, 1971). Local learning is controlled by the learning probability, learning rate, and the number of learning cycles.

2.4 SEA PERFORMANCE

The performance of the SEA model for the PCP problem has been studied in a series of experiments designed to investigate high-level behavioural patterns observed in the modelling with SEA, to establish the effective settings and interdependencies between SEA parameters, to examine the performance of spatial evolutionary algorithms on complex surfaces and planar covering problems, and to compare the performance of SEA with two non-evolutionary location algorithms.

Experiments designed to study the high-level search patterns employed by SEA and parameters dependencies used, as the demand surface, one composed of four squares touching at corners (Figure 2.2). Experiments designed to study the dependency between SEA performance and surface complexity used fractal surfaces representing Koch triangular and square islands and Sierpinsky negative random and deterministic carpets (Krzanowski, 1997). Finally, experiments studying SEA performance on complex location problems, and the experiments comparing SEA with non-evolutionary heuristics, used geographic objects including polygons representing the counties of Westchester, Suffolk, and Manhattan, the boundaries of Britain and the Antilles, and data sets representing road networks for rural, suburban and urban environments. In all the experiments, except for those using road networks, the demand was distributed evenly across the surface. In all of the experiments SEA modelled a planar covering problem with the same objectives: to cover the maximum demand with a given number of facilities of a constant radius. In each series of experiments different parameters settings were used and different metrics of SEA performance were recorded. The parameter settings have been selected to optimize the SEA performance for the particular tests; the metrics depended on the objectives of the particular experiment.

The series of experiments on the demand surface composed from four squares inspired a useful metaphor for evolutionary search: The search behaviour of SEA can be compared to the focusing of lenses. SEA search initially covers the whole search space, thereafter *focusing-in* upon select areas that provide the best objective function values. Figure 2.2 demonstrates this process. In this figure, the concentration of the search in specific areas is observed as the process of the concentration of objects (circles) at

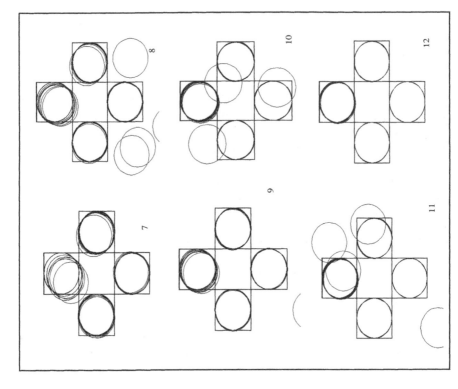

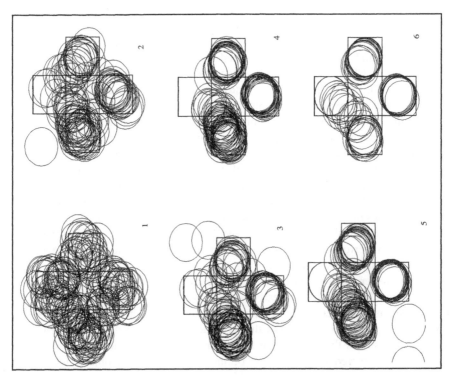

Figure 2.2 Location of objects during the evolution with Spatial Evolutionary Algorithm. The numbers below the illustrations refer to the iteration of the evolutionary cycles.

specific locations with outlines of circles becoming progressively more distinct. These experiments also established the influence of operators upon SEA performance. Contrary to the experience with non-spatial evolutionary methods, it was observed that mutation had no effect on SEA performance. The important SEA operators were learning (and its parameters), a type of cross-over, and the population size.

The experiments on different data sets and different problem complexities (defined as different number of facilities to locate) have demonstrated that SEA performance depends on the complexity and type of a data set – whether representing an artificial or a natural surface – as well as on the complexity of the PCP problem. However, this dependency diminishes as the problem or a surface become more complex. That is to say, in cases in which the number of facilities is large (in the experiments it was for the number of facilities greater than 35) or the surface is complex (those included complex fractals or non-simply-connected surfaces), the SEA performance is almost problem- or surface-independent. It was also observed that the SEA performance can change up to 50 per cent from a simple to a complex problem. To illustrate the capabilities of SEA to model a PCP problem, selected results of SEA runs on the geographic data set representing Britain and Suffolk County are given in Figure 2.3.

For the comparison of SEA performance two non-evolutionary heuristic algorithms have been chosen: Random placement algorithm (RPA) and Tornqvist's algorithm (TA). Random algorithms, such as RPA, are used in algorithmic analysis to establish minimum performance requirements for an algorithm. Performance at the level of random algorithms is a result of chance alone, not of organized action – i.e. a good algorithm should perform better than its random counterpart. TA is regarded as the best existing heuristics for planar covering problems (Tornqvist et al., 1971). It is a deterministic algorithm from a family of local hill-climbing search methods.

The tests comparing SEA, RA, and TA performance consisted of ten runs of spatial evolutionary algorithms for a planar covering problem on eight geographic data sets, 50 runs of RPA and 20 runs of TA on the same data sets and for the same definition of planar location problems. Results of those tests demonstrated that SEA performed up to 50 per cent better than RPA and five to ten per cent better than Tornqvist algorithm on all tested data sets. The results of tests on eight geographic data sets are presented in Table 2.1. The numbers in Table 2.1 are ratios of the SEA results to the results achieved by RPA or TA. The larger the number, the better SEA performance in the comparison to non-evolutionary heuristics. The columns entitled 'Best SEA' provide the ratio of the best SEA results to the best results achieved by RPA or TA respectively. The columns entitled 'Average SEA' provide the ratio of the average SEA results to average RPA and TA results.

2.5 CONCLUSIONS

Major findings and conclusions from the investigations conducted on SEA models of PCP are summarized as follows:

■ Spatial Evolutionary Search may be thought of as a process of focusing lenses: Initially SEA looks at the whole search space and then focuses on the optimal areas. What this metaphor is telling us is that SEA, contrary to traditional search methods which conduct the search from a single location, starts the search over the whole search space, and then narrows the search to those areas providing the best fitness.

BRITAIN

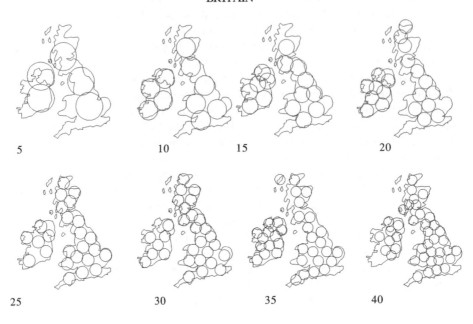

SUFFOLK

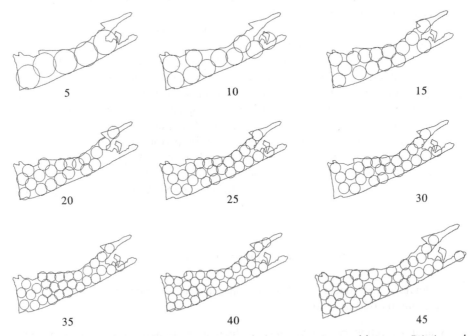

Figure 2.3 Results of tests of SEA on a series of planar covering problems on Britain and Suffolk data sets. Numbers below pictures represent the number of facilities in planar covering problem.

Table 2.1 Comparison of SEA performance with non-evolutionary heuristics on PCP.

Data set	Best SEA over Random	Best SEA over Tornqvist	Average SEA over Random	Average SEA over Tornqvist
Westch	1.36	1.09	1.50	1.13
Antyle	1.33	1.30	1.32	1.19
Suffolk	1.22	1.12	1.50	1.15
Manhat	1.34	1.13	1.63	1.15
Suburb	1.21	1.09	1.37	1.09
Rural	1.23	1.15	1.33	1.11
Britain	1.39	1.11	1.53	1.18
Urban	1.30	1.21	1.44	1.25

Suburb – suburban road network; Rural – rural road network; Urban – Urban road network

- The key SEA parameters are learning probability, learning rate, and population size. Parameters have an optimum setting beyond which SEA performance seems to stabilize: improper settings (too low) will affect SEA performance negatively.

- Spatial evolutionary algorithm performance is affected by the type of the complexity of the demand surface (fractal type, fractal dimension, and topology) and by the degree of the complexity of the surface. It has been shown that the most difficult surfaces are *not-simply-connected*. No strict dependencies between SEA performance and surface complexities have been observed or formulated.

- In all tested cases SEA performed significantly better than the random placement algorithm and much better than the Tornqvist algorithm. Those differences ranges were five to ten per cent superior to Tornqvist algorithms. In particular, SEA performed better on complex PCP over complex surfaces

This research into SEA also demonstrated that studies and research on the new modelling methods in spatial information systems (such as genetic programming, neural networks, data mining and knowledge discovery) should emphasize the specificity of spatial information and extensions of non-spatial methods required to handle spatial problems. In the particular case of the SEA presented in this study the specificity of the model has been incorporated into the algorithm through the hierarchical architecture of the evolutionary units mimicking the architecture of modelled spatial systems, through the introduction of fitness operators based on the map algebra, and the incorporation of the learning algorithm based on Tornqvist; none of those constructs are common or even defined for non-spatial EA.

References

BIANCHI, G. and CHURCH, R.L. (1992) *A Non-Binary Encoded Genetic Algorithm for a Facility Location Problem*, unpublished manuscript.

BROOKES, C.J. (1996) A Genetic Algorithm For Locating Optimal Sites on Raster Suitability Maps, *1st Int. Conf. on GeoComputation*, UK.

COOLEY, R.E. and HOBBS, M.H.W. (1992) An Application of AI to Computing Class Partition Values for Thematic Maps, in *Proceedings of 5th SDH*, Charleston, SC, 371–380.

DAVIS, L. (Ed.) (1991) *Handbook of Genetic Algorithms*, New York: Van Nostrand Reinhold.

DELAHAYE, D., ALLIOT, J-M., SCHOENAUER, M. and FARGES, J-L. (1994) Genetic Algorithms for Partitioning Air Apace, *10th Conf. on Artificial Intelligence for Applications*, 291–297. Los Alamitos: IEEE Computer Society Press.

DIBBLE, C. (1994) Beyond Data: Handling Spatial and Analytical Contexts with Genetic Based Machine Learning, in T.C. Waugh and R.G. Healey (Eds), *6th Int. Symp. on Spatial Data Handling*, 1041–1060, Edinburgh, UK.

DIBBLE, C. and DENSHAM, P.J. (1993) Generating Interesting Alternatives in GIS and SDSS Using Genetic Algorithms, *GIS/LIS Proceedings*, Minneapolis.

GOLDBERG, D.E. (1989) *Genetic Algorithms in Search, Optimization, and Machine Learning*, Readings, Addison-Wesley.

HOBBS, K. (1995) Spatial Clustering with a Genetic Algorithm, in *GISRUK 95*, UK.

HOBBS, M.H.W. (1993) A Genetic Algorithm For Knowledge Discovery in Spatial Data, in *GIS Research UK, First Nat. Conf.*, Keele, UK.

HOBBS, M.H.W. (1994) Analysis of a Retail Branch Network: A Problem of Catchment Areas, in *GISRUK 94*, Leicester, UK.

HOLLAND, J.H. (1993) *Adaptation in Natural and Artificial Systems* (2nd Edn), Cambridge, MA: MIT Press.

HOSAGE, C.M. and GOODCHILD, M.F. (1986) Discrete Space Location–Allocation Solutions from Genetic Algorithms, *Annals of Operations Research*, 35–46.

KELLER, S. (1995a) Potentials and Limitations of Artificial Intelligence Techniques Applied to Generalization, in J-C., Miller, R. Webel and Salge F., Lagrange, J.P. (Eds), *GIS and Generalization: Methodology and Practice*, London: Taylor & Francis.

KELLER, S.F. (1995b) Interactive Parameter Setting of Line Generalization Operators Using Genetic Algorithms, in *Proc. 17th Int. Cartographic Conf. (ICC'95)*, 3–6 Sept. 1995, Barcelona, Spain.

LIEPINS, G.E. and HILLARD, M.R. (1989) Genetic Algorithms: Foundations and Applications. *Annals of Operations Research*, **21**, 31–58.

MICHALEWICZ, Z. (1992) *Genetic Algorithms + Data Structures = Evolutionary Programs*, New York: Springer-Verlang.

MITCHELL, M. (1996) *An Introduction to Genetic Algorithms*, Cambridge, MA: MIT Press.

NISSEN, V. (1993) Evolutionary Algorithms In Management Science, *Papers on Economics and Evolution #9303*, Goettingen: University of Goettingen.

OPENSHAW, S. (1988) Building an Automated Modelling System To Explore a Universe of Spatial Interaction Models, *Geographical Analysis*, **20** (1).

OPENSHAW, S. (1992) Some Suggestions Concerning the Development of Artificial Intelligence Tools for Spatial Modeling and Analysis in GIS, *Annals of Regional Science*, **26**, 35–51.

OPENSHAW, S. and TURTON, I. (1996) Building New Spatial Interaction Models Using Genetic Programming, unpublished manuscript.

PEREIRA, A.G., PECKAM, R.J. and ANTUNUS, M.P. (1996) GENET: A Method to Generate Alternatives for Facilities Siting Using Genetic Algorithms, *EuroGIS*, 1996.

RAPER, J.F. (1989) Key 3D Modelling Concepts for Geoscientific Analysis, in Turner, A.K. (Ed.), *Three-Dimensional Modeling with Geoscientific Information Systems*, London: Kluwer Academic Publishers.

RAPER, J.F. and LEVINGSTON, D. (1997) *Spatio-Temporal Interpolation in Four Dimensional Coastal Process Models*, GIS UK 97.

TOMLIN, D.C. (1985) *Map Algebra*, Cambridge, MA: Harvard University.

TORNQVIST, G., NORDBECK, S., RYSTEDT, B. and GOULD, P. (1971) Multiple Location Analysis, *Lund Studies in Geography*, No. 12.

CHAPTER THREE

Computational techniques for non-Euclidean planar spatial data applied to migrant flows

MICHAEL WORBOYS, KEITH MASON AND JENNY LINGHAM

3.1 INTRODUCTION

Almost all current GIS functionality assumes that constituent spatial entities in the system are referenced to the coordinatized Euclidean plane – that is, they all have coordinates positioning them with respect to a predetermined reference frame. In addition, the usual Euclidean properties and relationships, such as distance and bearing between locations, lengths and angles of line segments, and area and perimeter of regions, may be calculated in the normal way. However, there are many situations for which representation in a Euclidean space is not the most appropriate model, nor may even be derivable. Examples include travel time spaces, qualitative distances and flow spaces.

This paper considers the potential for spatial analysis that takes us outside the geometry associated with Euclidean space. Topology is the key concept, and the paper explores the potential for using topological concepts, even when a metric or Euclidean space is not available. In particular, since many geospatial phenomena are best modelled in a finite context, we explore the potential of topologies on finite sets (so-called *finite topologies*). We show that finite topologies provide an analysis of collections of spatial entities as hierarchies.

The methods are applied to flow spaces, resulting from linkages between geographically-distinct areas brought about by 'flows' between origins and destinations, and exemplified by journeys-to-work, diffusion of disease or of innovations, population migrations, freight flows, international trade, patient referrals to hospital and shopping patterns. In seeking to represent or analyze such interactions, the objective is to understand or to model and predict the flows, and a variety of techniques for such analysis and visualization are presented in this paper.

3.2 BACKGROUND

3.2.1 Visualizing interactions: flow mapping

Of all cartographic forms, flow mapping is one of the most problematical. Perhaps the simplest visualizations are those that deal only with the end product – choroplethic representations showing rates of inflow or outflow per unit area, without any indication of respective origins or destinations. Alternatively, symbolization of the interaction itself, the flow, may be used whenever visualization of dynamic linkages between places is required. Broadly, flow maps are constructed according to one of two basic styles: those which are concerned with qualitative movements and which symbolize linkages and directionality by way of simple unscaled lines and arrowheads, and those which are concerned also with quantitative movements, where the width of the line is varied in proportion to the amount of flow recorded between places.

A number of innovative variations upon these solutions to mapping flows have also been reported, for the most part in avoidance of problems associated with line overlap and visual clutter. One widely used variant (see, for example, Dorling, 1991) is that which symbolizes connectivity through the use of flow lines of equal width, but also portrays quantity of movement by means of proportional symbols placed at line ends or at the centre of areas receiving or generating the flows.

3.2.2 Analyzing interactions

Moving beyond visualization, spatial interaction models allow investigation of the determinants of flow spaces. Spatial interaction models usually operate on the premise that flows are likely to be in some way related to the sizes of the origin and destination areas, and in some way inversely related to the distance between them. By such means, patterns of flows can be modelled in terms of accessibility of destinations from origins, and concepts of demand and attractiveness – the ability of different areas to generate a flow – can be introduced into the analysis. One of the most common classes of spatial interaction model is the gravity model (Wilson, 1974), where flow is analogous to gravitational attraction, and the generating and pull factors associated with origins and destinations are analogous to mass. Where the fitted model exactly reproduces the total observed flow, such models are referred to as being 'doubly constrained' – separate parameters being allowed for each origin and destination. However, whilst the doubly-constrained model provides useful insights into the role of distance in encouraging spatial interaction, and information on the relative contributions of origins and destinations to the process, it offers little in the way of explanation for the observed flows. Consequently, a number of variations upon the basic model might be employed. The 'origin-constrained' model attempts only to ensure that the predicted flows from an origin equal the actual flow, while the 'destination-constrained' model is concerned that the reverse is true – that predicted flows into a destination are equal to those actually observed. The advantages of both adaptations is that they allow alternative configurations or additional origin or destination parameters to be incorporated into the model to predict total flows.

The overall patterns of flow associated with a particular place can be summarized through the use of migration fields. A migration field delineates the origins of immigrants to an area or the destinations of emigrants from that area. Typically, the fields will be defined as those areas contributing more than a specified number of migrants to the overall pattern of flow. A technique known as primary linkage analysis can be used to identify such fields (Haggett, Cliff and Frey, 1977). In this, the hierarchical order of a region is measured by the number of people moving to that region, and the largest outflow to a higher order region is used to determine the overall hierarchical structure.

3.2.3 New approaches to analysis and visualization using non-Euclidean spaces

In this paper we investigate ways of representing and reasoning about interactions between spatial objects that are not necessarily referenced to the Euclidean plane. The work is motivated by the interest within the research community in the idea of 'geographic space', that may not be Euclidean (see for example the concept of 'naive geography' introduced by Egenhofer and Mark, 1995). Our view of the geographic space which surrounds us may be quite different from the representation of that space on a map; the distance relationship may not be symmetric, in that it may be quicker to go from A to B than it is to go from B to A, perhaps because of the terrain or the traffic conditions. We can think of ourselves as being immersed within geographic space, unable to view the whole of it at any one time. In order to explore it we must move around in it, both spatially and temporally. This space is thought of as being horizontal and two-dimensional. It emphasizes topological information, lacks the usual distance metric, and frequently lacks complete information.

Topology is the study of the properties of a space which are preserved when it is subjected to continuous transformations – for example, rotations or stretches. For a general treatment of topology the reader is referred to Kuratowski (1996) or to Henle (1979). Digital topology has been used in fields such as computer vision, graphics and image processing for the study of the topological properties of image arrays. There is a substantial research literature on this subject, for instance Kong and Rosenfeld (1989). The work discussed here uses finite topologies, the topologies of finite sets (see for example Kovalevsky, 1989). As topology is essentially the study of continuity and limiting processes, it might be thought that finite topology would provide us with little insight. However, finite topologies do provide a hierarchical structure on sets of entities, which is precisely the structure which we develop later in the paper. A start to this line of research is described by Worboys (1996) and defines a topology in a non-metric space, based on the construction of localities. Related work has been carried out by Egenhofer and Franzosa (1991) who present a framework for the definition of topological relations which is independent of the existence of a distance function.

Montello (1993) has discussed the importance of alternative geometries in the study of the large-scale space in which we live, and which he refers to as 'environmental space'. Montello notes that movement over time is usually required to explore this space, and that many authors have suggested that a Euclidean model is not the best one to represent it. Gahegan (1995) proposes qualitative spatial reasoning as a way of reasoning about objects which deals with topological relationships between them rather than their positions in space. It is suggested that the idea of proximity need not imply physical closeness, and could also depend on the context.

Another approach to the problem of dealing with objects in non-Euclidean space has been the use of multi-dimensional scaling (MDS). This is described by Gatrell (1983). MDS is used when we have a set of objects and wish to create a map representing the relationships between the objects. The input to a multi-dimensional scaling algorithm is a matrix of dissimilarities or distances. The output is a spatial configuration of the objects, so that the distances between pairs of objects within the configuration match as closely as possible the input dissimilarities. The effect of the MDS is to distort the Euclidean plane. Gatrell also describes and gives examples of the use of Q-analysis, based on the work of Atkin (1978). Q-analysis is a technique that can reveal connectivities provided by relations on a set, or between sets, using concepts from algebraic topology.

Other important work which does not assume a Euclidean framework but attempts to describe directly spatial relationships such as contact and boundary includes the work of Clarke (1981, 1985), and later Cohn *et al.* (1993) on the Region Connection Calculus (RCC). Smith (1995) gives an account of mereotopology, which combines mereology (study of the part–whole relationship) and topological notions.

3.3 MIGRANT FLOWS

3.3.1 Population migration

In the UK, studies of migrant flows have received much recent attention, largely as a result of the role of internal migration in shaping local population distributions and structures, in circumstances of little or no natural population increase. Several important trends in population migration in the UK have been observed at different times and at a variety of geographical scales. Long-distance movements related to place-of-work have been particularly significant. Rural to urban and then, through counter-urbanization, urban to rural population movements have been instrumental in population mixing across the country.

Movements between Family Health Service Authorities (FHSAs) within England and Wales during the 1980s have been measured by the National Health Service Central Register (Rosenbaum and Bailey, 1991). By using patient reregistrations, the biggest net gainers of population through migration, in terms of average annual gain, were found in the South West, the South East (excluding Greater London), and East Anglia. The principal losers were in Greater London. In broad terms, it was the non-metropolitan FHSAs which gained population during the 1980s, and the metropolitan FHSAs which lost. In Scotland, migration flows have been calculated from the Register of Sasines, which records every sale of *inter alia* private residential property (McCleery, Forbes and Forster, 1996). From this, short-distance migration is seen to dominate – a factor which is sometimes blurred when more coarse data has been employed elsewhere.

There are several reasons why population movements deserve close attention. Clearly, as migration is an important component of population change, a proper understanding of it is required to construct realistic population projections and to inform resource allocation decisions. In addition, the selective nature of migration can shape and re-shape the social and demographic characteristics of those areas losing or gaining populations by such means. This has implications for local authority revenue raising and service provision. Similarly, land-use planning policies and levels of investment in housing are influenced by migration. At the same time, population movements have important implications for epidemiological studies, not least in the spread of disease through increased contact made between susceptibles and infectives, but also through the combined effects of migration and disease latency as a confounding problem in trying to explain patterns of disease distribution. The explanation for the occurrence of a disease will have as much to do with past histories as it does with present sets of circumstances.

Given the significance of migration for shaping the geography of the United Kingdom, it is important that the range of techniques for visualizing and analyzing migrant flows is

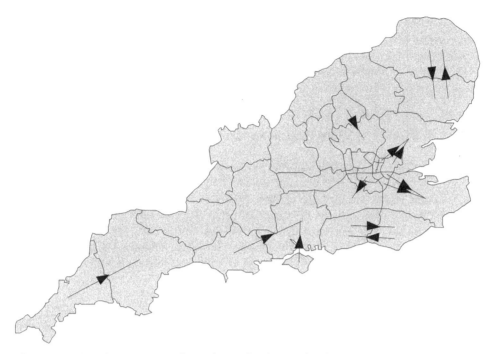

Figure 3.1 Significant migrant flows shown for the South of England.

as extensive as possible. The method developed below introduces an approach to analysis that is less dependent upon the Euclidean characteristics of the spatial framework.

3.3.2 The data

For present purposes the Special Migration Statistics (SMS) of the 1991 Census (Flowerdew and Green, 1993) will be used to establish migration flows between Family Health Service Authority (FHSA) areas. The latter have been selected for compatibility and comparability with the only other major dataset on migrant flows in this country, that developed from the National Health Service Central Register, which is also aggregated to FHSAs. In the SMS a migrant is defined as: 'a person with a different usual address one year ago to that at the time of the Census'. The SMS contain simple origin-destination flow counts as well as more complex cross-classified counts. Clearly, the SMS can provide nothing more than a snapshot of migration patterns as they existed in the year prior to the Census. Nevertheless, they are sufficient to allow explorations of the ways in which non-Euclidean approaches may be brought to bear, and at the same time throw new light on spatial interactions as they exist through migrant flows in the UK. Figure 3.1 shows a representation of the dataset for the South of England, where an arrow is shown between region A and region B if not less than 20% of the total flow from A goes to B.

3.4 THE NON-EUCLIDEAN SPACE OF LOCALITIES

3.4.1 From flows to proximities

The technique will be explained with reference to the migration flow data described earlier. It should be emphasized that the method is quite general and may be applied to any proximity or accessibility relationship between spatial entities. The migration flow data consists of a 113 by 113 matrix $\mathbf{F} = [f_{ij}]$ of migratory flows, where f_{ij} is an integer representing the amount of flow from location i to location j, the locations being the 113 FHSA regions in England, Scotland and Wales. We convert the flows into normalized measures of proximity (or accessibility) p_{ij} by means of the functional relationship

$$\log_2(1 - p_{ij}) = -f_{ji}/\mu$$

where μ is the mean of all the flows in the matrix. For our example, $f_{12} = 1867$ and so $p_{21} = 0.97$.

Note that $0 \leq p < 1$, small flows result in near-zero proximity values, large flows result in values of proximity near to one, and a mean flow results in a proximity value of 0.5. The intuition is that the quantity of flow from j to i is an indication of the level of accessibility of i from j or the proximity of i to j. This measure of proximity is very different from the usual Euclidean distance – not necessarily being symmetric, for example. We are thus unable to proceed with the usual Euclidean analysis. The next step in our approach is to convert the proximity measure to a Boolean relation π with reference to a fixed cut-off parameter κ as follows:

$$i\pi j \text{ if and only if } p_{ij} > \kappa \tag{3.1}$$

In our example, we take $\kappa = 0.6$, and as $p_{21} = 0.97$, it is true that $2\pi 1$. We assume that $i\pi i$ for each location i, thus π is a reflexive relation. This assumption is strengthened by the large values observed for flows from a location to itself.

3.4.2   Localization

The next subsection will show that to derive a topology from the proximity relationship, it must be a partial order. A relation ρ is a partial order if it satisfies the following properties:

For all locations i,	$i\rho i$	REFLEXIVE
For all locations i, j, k	$i\rho j$ and $j\rho k$ implies $i\rho k$	TRANSITIVE
For all locations i, j	$i\rho j$ and $j\rho i$ implies $i = j$	ANTISYMMETRIC

We already have assumed that the proximity relation is reflexive. However, it is unlikely that any nearness type relation will be transitive. The antisymmetric rule holds if there do not exist distinct locations i and j for which i is proximal to j and j is proximal to i, again unlikely to hold.

We get the properties of a partial order from our proximity relation by a localization process now to be described. Define a new relation λ on the set of locations as follows. For all locations i, j, $i\lambda j$ if and only if any location which is proximal to i is proximal to j. That is:

$$i\lambda j \text{ if and only if for all locations } k, k\pi i \text{ implies } k\pi j \tag{3.2}$$

We read $i\lambda j$ as 'location i is local to location j'.

It is not hard to show that λ is both reflexive and transitive. It may also be made antisymmetric by factoring out (coalescing) all locations i, j for which $i\lambda j$ and $j\lambda i$. So relation λ provides a partial order on the set of locations, thus providing a hierarchy of locations.

The partial order for the migratory flow data is shown in Figure 3.2, where a directed segment from i to j represents the relation $i\lambda j$. In Figure 3.2, the visualization relies on a

Figure 3.2 Locality relationship shown for the South of England.

mix of Euclidean and non-Euclidean spaces, the embedding of locations being Euclidean but the partial order being non-Euclidean.

3.4.3   Finite topologies

Many fundamental relationships between spatial entities are describable in topological terms. A spatial relationship is *topological* if is invariant under continuous deformations of space. The topology that has been of almost universal concern in geospatial data handling is the so-called 'usual topology' of the Euclidean plane, derived from the 'as-the-crow-flies' distance function. But many other topologies exist, in all of which such concepts as interior, boundary and connectivity are defined and may have useful interpretations. In the general setting, a topological space is a set with a distinguished collection of subsets (called the *open sets*). These open sets must obey the following *three* conditions.

- ■   T1   The empty set and the entire space are open sets.
- ■   T2   The union of any family of open sets is an open set.
- ■   T3   The intersection of any finite family of open sets is an open set.

Also, if the topological space satisfies the further property that, for any two distinct points in the space, there is an open set which encloses one but not the other, then the space is called *separable*.

The topologies that we use in this work are finite, in the sense that the underlying set is finite. It might be argued that restricting consideration to finite topologies misses many important aspects of topology, including continuity and limit processes. However, many models of real world data and processes are finite, and we contend that for finite topologies useful properties remain. Condition T3 implies that the intersection of all collections of open sets in a finite topology is open. For any location, consider all open sets that have it as a member, then the intersection of these open sets will be open, and is the minimal open set to which it belongs. Minimal open sets do not necessarily exist in the infinite case.

The key result for this work is that every separable, finite topology may be conceived as a partially ordered set. And, conversely, any finite partially ordered set may be structured as a separable, finite topology. The proof of this result relies heavily on the concept of minimal open sets and is not hard, although outside the scope of this paper, and may be found in Worboys (1997).

Returning to our flow analysis, at the end of the last subsection we showed that the locality relation λ results in a partial order on the set of FHSAs, and we have just shown that this leads to a topology. The remaining questions are how can this topology be visualized, and how does it help us in analysis of migration between FHSAs.

3.4.4   Visualization of finite topologies

This sub-section presents two separate visualizations of a finite topology (and therefore a finite partially-ordered set). The first approach is to use the notion of a topological neighbourhood of a location. In the finite case, the minimal neighbourhoods of locations exist and correspond to the minimal open sets. Intuitively, these neighbourhoods may be thought of as regions of influence of their locations. In the case of proximity neighbourhoods (that is, neighbourhoods arising from the localization of proximity relations) the neighbourhoods may be thought of localities. Formal details of their properties may be found in Worboys (1997). Figure 3.3 shows the neighbourhoods in the South of England in the case of the FHSA data.

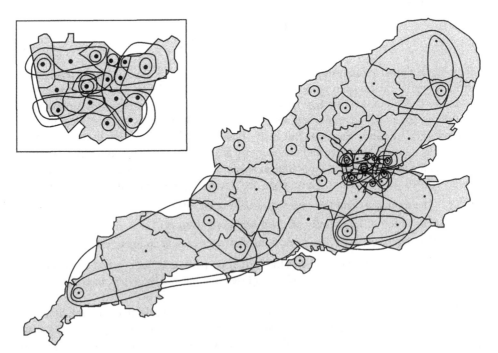

Figure 3.3   Neighbourhoods shown for the South of England with inset showing expanded view of London.

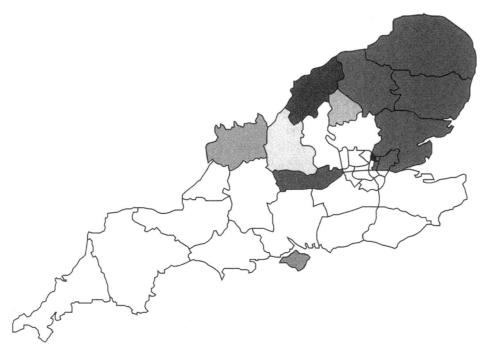

Figure 3.4   Partitioning localities for the South of England.

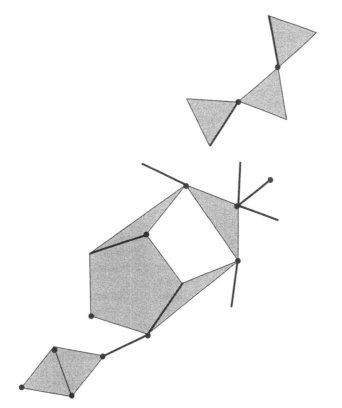

Figure 3.5   The FHSA migratory flow data for the South visualized as a cellular complex.

These neighbourhoods can be combined into maximal connected components of the space. The resulting zoning provides some insight into the partition of the space into localities, each of which is a relatively self-contained region of spatial interaction (in this case through migration). Figure 3.4 shows the partition of the South of England.

The second approach relies on the result (again not proved here, see Worboys, 1997 for details) that the hierarchy relating to a finite topology may be visualized as a cellular complex. A *cell* can be thought of as a face of a *n*-dimensional polytope: a point in 0-dimensions, straight line segment in 1-dimension, polygon in 2-dimensions, polyhedron in 3-dimensions, and so on. A cellular complex is a collection of cells, with the property that any face of a cell in the complex must also be in the complex. A hierarchy relating to a finite topology may be visualized as a cellular complex, with locations down the hierarchy represented as lower-dimensional faces of locations higher in the hierarchy. Figure 3.5 shows part of the cellular complex that represents the FHSA example.

References

ATKIN, R.H. (1978) Time as a pattern on a multi-dimensional structure, *J. Social Biol. Struct*, **1**, 281–295.

CLARKE, B.L. (1981) A calculus of individuals based on 'connection', *Notre Dame J. Formal Logic*, **22**, 204–218.

CLARKE, B.L. (1985) Individuals and points, *Notre Dame J. Formal Logic*, **26**, 61–75.

COHN, A.G., RANDELL, D.A., CUI, Z. and BENNETT, B. (1993) Qualitative Spatial Reasoning and Representation, in P. Carrete and M.G. Singh (Eds), *Qualitative Reasoning and Decision Technologies*, CIMNE, Barcelona.

DORLING, D. (1995) Visualizing the 1991 census, in S. Openshaw (Ed.), *Census Users Handbook*, Cambridge: GeoInformation International.

EGENHOFER, M.J. and FRANZOSA, R.D. (1991) Point-set topological spatial relations, *Int. J. Geographical Information Systems*, **5** (2), 161–174.

EGENHOFER, M.J. and MARK, D.M. (1995) Naive geography, *Technical report 95–8*, National Centre for Geographic Information and Analysis, Santa Barbara: University of California.

FLOWERDEW, R. and GREEN, A. (1993) Migration, transport and workplace statistics from the 1991 Census, in A. Dale and C. Marsh (Eds), *The 1991 Census Users Guide*, HMSO.

GAHEGAN, M. (1995) Proximity operators for qualitative spatial reasoning, in A.U. Frank, and I. Campari (Eds) *Spatial Information Theory, Proceedings of COSIT 95, Lecture Notes in Computer Science*, Berlin: Springer-Verlag, 31–44.

GATRELL, A.C. (1983) *Distance and space: A geographical perspective*, Oxford: Clarendon Press.

HAGGETT, P., CLIFF, A.D. and FREY, A. (1977) *Locational Analysis in Human Geography*, London: Edward Arnold.

HAYES, P. (1978) The Naive Physics Manifesto, in D. Mitchie (Ed.), *Expert Systems in the Microelectronic Age*, Edinburgh: Edinburgh University Press, 242–270.

HENLE, M. (1979) *A Combinatorial Introduction to Topology*, San Francisco: Freeman.

KONG, T.Y. and ROSENFELD, A. Digital Topology: Introduction and Survey, *Computer Vision, Graphics and Image Processing*, **48**, 357–393.

KOVALEVSKY, V.A. (1989) Finite topology applied to image analysis, *Computer Vision, Graphics and Image Processing*, **46**, 141–161.

KURATOWSKI (1996) *Topology*, New York: Academic Press.

McCLEERY, A., FORBES, J. and FORSTER, E. (1996) Deciding to move home: A preliminary analysis of household migration behaviour in Scotland, *Scottish Geographical Magazine*, **112** (3), 158–168.

MONTELLO, D.R. (1992) The geometry of environmental knowledge, in A.U. Frank, I. Campari and U. Formentini (Eds), *Theories and Methods of Spatio-Temporal Reasoning in Geographic Space, Lecture Notes in Computer Science 639*, Berlin: Springer-Verlag. 136–152.

ROSENBAUM, M. and BAILEY, J. (1991) Movement within England and Wales during the 1980s, as measured by the NHS Central Register, *Population Trends*, **65**, 24–34.

SMITH, B. (1996) Mereotopology: a theory of parts and boundaries, submitted to *Data & Knowledge Engineering*.

WILSON, A.G. (1974) *Urban and Regional Models in Geography and Planning*, London: Wiley.

WORBOYS, M.F. (1996) Metrics and Topologies for Geographic Space, in M.J. Kraak and M. Molenaar (Eds), *Advances in Geographic Information Systems Research II: Pro. Int. Symp. on Spatial Data Handling, Delft*, International Geographical Union, 7A.1–7A.11.

WORBOYS, M.F. (1997) Using finite topologies to extract hierarchies from data referenced to non-Euclidean spaces, submitted for publication.

CHAPTER FOUR

New interactive graphics tools for exploratory analysis of spatial data

ANTHONY UNWIN AND HEIKE HOFMANN

4.1 INTRODUCTION

Geographic Information Systems organize vast amounts of data, but that does not mean that all the information they contain is easy to access. You have to know what you are looking for and you need tools which enable you to search the data effectively. There are two main alternatives: setting automatic analytic procedures loose on the data to see what they come up with (an extreme version of data mining); and using exploratory data analysis. Obviously, the most successful approach will be a judicious mixture of the two, using exploratory methods to guide where and how data mining should be employed, and using data mining to generate subsets of information of interest to be investigated in depth with exploratory methods. Our work concentrates on the introduction and development of new graphical tools for exploratory analyses.

Exploratory methods are an important means of dealing with spatial data, as the assumptions for traditional confirmatory statistical methods no longer hold: points (or areas) are not independent, distributions are often highly skewed, and special geographic features such as borders and rivers must be taken into account. Unfortunately, the term 'Exploratory Data Analysis' (EDA) in statistics has been closely associated with the methods described in Tukey's important book of 20 years ago (1977). This is really now more of historical interest, as computing power has developed so dramatically since then. Much more effective methods are available, particularly interactive graphical ones, which implement the philosophy of EDA implicit in Tukey's book, although with sophisticated software rather than with the mainly pencil and paper approaches he described (e.g. Becker *et al.*, 1987; Unwin, 1994; Velleman, 1994). Graphical methods are especially valuable for the exploration of geographical data, and these applications are a source of interesting new problems for statisticians. It is also instructive for statisticians used to working with non-spatial data to see how geographers have tackled the problems of visualizing spatial information.

4.2 EXPLORATORY DATA ANALYSIS

Exploratory analysis may be used for a range of goals. It is excellent for examining data quality. Values which can hide in tables stand out in graphics, especially when linked to further displays such as maps. Of course, it is important to check whether a value is an error or a real, but extreme, value. Linking to displays of other variables is valuable for assessing the internal consistency of an outlier. There is a simple but unusual example in the Mid-West data set. If the percentage of black population is plotted against the percentage of white population for the counties, there is a very strong negative linear association and an extreme outlier at close to (0,0). Is this an error or a genuine, but exceptional, value? Linking to the percentages of the Asian-Pacific, American Indian/Eskimos and other races shows that there is indeed a county with a predominantly Indian population, which explains the low values for white and black percentages.

Assuming that the quality of the data has been checked and that the most extreme outliers have been identified – because outliers are only outliers in relation to an implicit understanding of what is a 'standard' range of values, new outliers may be identified in a continual process, as analysts' views of what is standard become more and more refined – further goals are to identify groups or clusters of points and possible structure in the relationships between variables. The central importance of geographic clustering in spatial data and the wide variety of patterns which may arise make a visual approach essential. While in an ideal world we would wish to be able to assess the statistical significance of what we observe, the range of results of value would be severely curtailed if we insisted on this. Areas along a border might form a group, or areas around, but not including, cities, or rural areas in the North, or . . . A statistical test could be devised for each one of these patterns in advance, but it is not known in advance which patterns might appear. Instead, as with examining outliers, it is necessary to consider the implications of any apparent result for the other sources of information available, whether in terms of quantitative or qualitative variables or of geographic features. Are the areas along the border different because people living close by can easily cross it to shop? Do the areas round the cities form a group because they are all dormitory towns for the city workers? Are agricultural practices different in the North because of the climate and the type of land? This kind of assessment of results is very much in line with good scientific practice: always seek external evidence from other sources which could disprove your hypotheses. 'External' here is taken to mean data that have not been used to generate the hypotheses. Traditional confirmatory statistical analysis applied as part of EDA is more concerned with the weight of internal evidence to support a hypothesis which was first generated from the same data.

While it may not always be possible to check ideas generated from exploratory analyses using statistical modelling, results must be viewed with great caution without such support. Conversely, any results derived from analytic models should always be checked using exploratory methods. Interactive graphical methods are excellent for examining residuals and assessing fit qualitatively. The two approaches complement each other very well.

All exploratory analyses benefit from the interactive graphical approach. The concepts of direct interrogation and of linking in all its forms will be taken as given, although they are not yet as standard in statistical and other software as we would wish. What is of interest in this paper is to consider tools which are specifically designed for dealing with spatial data and to examine them in the general context of exploratory spatial data analysis. Many ideas were developed in the REGARD software written by Graham Wills (a description may be found in Unwin, 1994) and will not be further described here. (Applications and discussion can be found in several papers – Unwin et al., 1990; Wills et al., 1992; Unwin, 1995.) Instead we shall concentrate on new tools which have been introduced in MANET and what they can contribute to an overall approach to spatial analysis. The philosophy underlying both packages has been that it must be possible to view spatial data in a variety of statistical displays, that it is important to be able to view many variables simultaneously, and that it is crucial to be able to link all of these to one another and to geographic displays.

4.2.1 Interactive choropleth maps

Choropleth maps are a traditional tool for presenting spatial distributions by area. Each region is shaded, coloured or hatched according to the value associated with it. The literature includes many discussions of what makes a good choropleth map, and the difficulties associated with displaying data in this static way. MANET implements interactive choropleth maps. Apart from the basic interactive tool of interrogation (so that area values and names can be directly queried from the map on screen with the mouse) and linking (so that areas can be selected and linked to other statistical displays), MANET offers direct controls for flexible rescaling of the shading through a functional mapping from the values to the shading scale.

The default scaling is that the range of observed values for the variable X $[x_{min}, x_{max}]$ is divided into 32 equal sections and the shading y goes from dark (= 0%) in the first section at min to light (= 100%) in the top section at max. That is, y is a step approximation to a linear function:

$$f(x) = (100/32) \times (x - x_{min})/d \tag{4.1}$$

where $d = (x_{max} - x_{min})/32$.

In Figure 4.1 there are a few areas with relatively low rates in the centre, but there appear to be clusters on the Southern borders and some low rates to the North. This is the default display produced by MANET, but it can be changed using a dialog box which offers several control options and displays how the distribution of values by shading changes in response. The range is divided into 32 equal intervals, marked by black lines and dots along a value axis. (It is planned to change this to allow the user to choose the limits of the range or to force common ranges for a set of maps, which would be particularly useful for comparisons, say of data from successive years.) The black lines are proportional to the numbers of cases with that shading and so give a coarse picture of the distribution of the data. Black dots are drawn for intervals where there are no cases. Red lines are drawn on the shading pattern for intervals with cases to give an overview of which shading levels are in use and the differences between them. There is an option to invert the shading (switching from $f(x) = x^b$ to $f(x) = 1 - x^b$) and an option to change to a non-linear form by altering the value of b. With the first function, making b larger distinguishes the higher values more, while making b smaller distinguishes the lower values more. Currently in development is a transformation which can emphasize differences in the centre or at both ends simultaneously using an approximation to an S-shaped function:

$$f(x) = a \times (x/a)^b \qquad\qquad x \leq a$$

$$f(x) = 1 - (1 - a) \times ((1 - x)/(1 - a))^b \quad x \geq a \tag{4.2}$$

where a determines the centring and b the form.

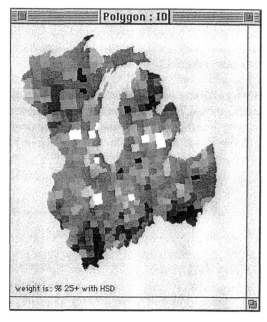

Figure 4.1   The 437 counties in the six states of the American mid-West, shaded according to the percentage with a High School Diploma: dark means low and light means high.

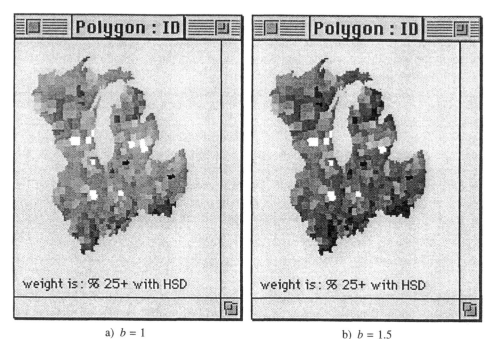

a) $b = 1$
Display a) highlights the counties with lower rates in the centre.

b) $b = 1.5$
Display b) highlights the clusters of lower rates in the North and South.

Figure 4.2   Choropleth maps of percentages of high school diploma.

As multiple linked maps can be displayed simultaneously, it is possible to compare both different shadings of the same variable and also shadings for different variables. The former is valuable for extracting different aspects of the information carried in the spatial distribution of a variable (see Figure 4.2).

Drawing a set of choropleth maps for different variables is useful for looking for spatial associations, although linking seems more effective for checking the strength of the patterns. With visual displays it is unwise to be too dogmatic, as features may become apparent in one display which were not at first sight visible in another. A key advantage of software like MANET is its ability to display many different views in parallel and pursue several complementary approaches simultaneously.

4.2.2 Weighted histograms

Regional data is often expressed as percentages, so that information can be comparable for regions of different sizes or populations. Histograms of these percentage distributions linked to the map are useful for understanding variability, but need to be interpreted with the absolute numbers in mind. An obvious example arises in plotting the distribution of disease rates. One way to do this is to link to plots of the raw data, but a new alternative is to use weighted displays. Instead of giving each region the same area in the histogram, another variable is chosen as a weighting, and the region is represented by an area proportional to this variable. For instance, a region with a population of one million gets ten times the area of a region of 100 000.

Histograms are commonly drawn with equal interval widths, so if the weight variable is W with value w_i at point i, the height of the bar for the interval $(a_j, a_j + 1)$ is proportional to Σw_i and the sum is overall i for which $a_j \leq x < a_j + 1$. If only some points in the bar are selected, the bar is highlighted to a height proportional to the sum of their weights.

Interrogating the bar furthest to the right in the weighted histogram in Figure 4.3 shows that 17.6% of the total Black population in these States live in areas with over 40% Blacks, but that this is only one county out of the 437. The range of possibilities opened up by this new graphic is surprisingly large. There are often good reasons for investigating several weighting variables for the same displayed variable as in Figure 4.3 for the percentages of an ethnic minority amongst the inhabitants of a region. You could weight by total population (the obvious default), by the absolute numbers of the ethnic group, by the numbers of other groups, and possibly by other variables as well. The only restriction is that the weighting variable should add naturally across the cases.

4.2.3 Local statistics

Recent collaboration between the statistical software group in Augsburg and geographers in Leicester and at Birkbeck College in London has highlighted the common interest of geographers and statisticians in examining local rather than global statistics. Global statistics depend on assumptions of comparability across the whole area covered by the data set and assume that the boundary is an immutable given. Local statistics offer a much more flexible view (which in principle encompasses global statistics in the limit). While the need for local statistics is clear, this does not mean that suitable analytic methods are available. Recent developments include LISA statistics – Local Indicators of Spatial

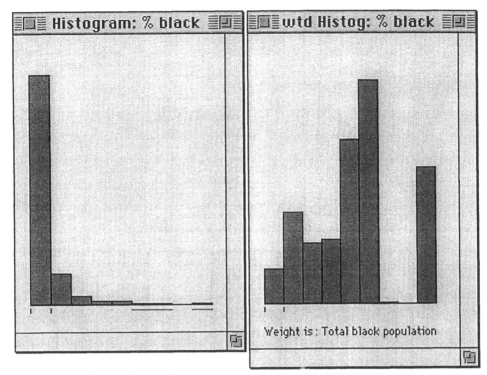

With no weighting, the display shows the numbers of counties within a particular percentage range of blacks. It is clear that most counties have low percentages of blacks.

Weighting by black population shows the numbers of blacks living in counties within a particular percentage range of blacks. Here we can see that most blacks live in counties with relatively higher percentages of blacks.

x axis scaling and histogram bin-widths are the same in both plots: 0% to 45% in 5% steps. The vertical axis scaling is 0 to 400 counties in the left hand plot, and 0 to 1.5 million black population in the right hand plot.

Figure 4.3 Percentage of black population weighted in different ways.

Association (Anselin, 1995) – and the G and G* statistics (Ord & Getis, 1995), but neither of these are easy to interpret and, more importantly, they are solely univariate, not multivariate. Interactive graphical tools are a valuable tool to be used in parallel with local statistics and have the advantage of being naturally multivariate. (See Wilhelm and Steck (1997) for further discussion and evaluation.)

A first and simpler step in this direction is to display dotplots or boxplots of the data for several variables and use linking between these displays and the map display to identify anomalous areas or interesting clusters. (In principle this could also be done for the local statistics themselves.) Both dotplots and boxplots make very efficient use of screen real estate so that drawing displays for as many as twelve variables in parallel is not unreasonable. (Contrast this with using histograms or scatterplot matrices.) Dotplots are useful for when there are not many cases, and also for identifying gaps in the data and hence possible clusterings. MANET has a special brightness control to allow dotplots to be drawn for larger data sets (when multiple values arise) but, in general, boxplots are better for large data sets. They are especially good for emphasizing outliers (since the

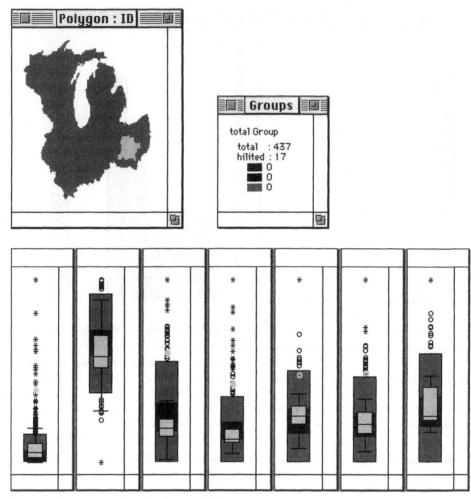

The seven boxplots are in order from the left:
% black (1 local outlier and 1 global outlier which is not local)
% over-25 with High School Diploma (1 global outlier which is not local)
% over-25 with College degree (1 local outlier which is also global)
% over-25 with a higher professional qualification (1 local outlier which is also global)
% aged 0–17 below poverty level (1 local outlier which is not global)
% aged 18–59 below poverty level (1 local outlier which is also global)
% 60 and over below poverty level (no outliers).

Figure 4.4 Linked boxplots for spatial data (N.B. These plots are not on a common scale).

extreme values are of most relevance in highlighting anomalies). In MANET boxplots for the whole data set are drawn in the background with shading out to the inner fences, while boxplots for the subset of highlighted points are drawn in the traditional box and whisker form in the foreground. Outliers for the whole data set will obviously be outliers for most subsets and will be found in most graphics. Local outliers show up as outliers only in the highlighted boxplots. Figure 4.4 shows the selection of 17 areas around Columbus, Ohio. Most of the univariate local outliers identified are also global outliers, but one is not, and one global outlier is not a local outlier.

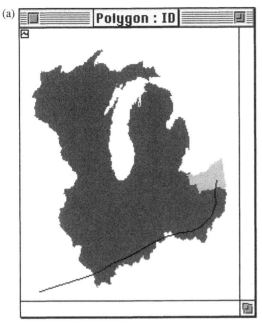

(a)

(a) The brush has been moved along the path shown in the map window. The neighbourhood counties for which the last set of statistics has been calculated are highlighted. The brush was set to calculate running means for the percentages of those below poverty level in the three age groups: 0–17, 18–59 and 60+.

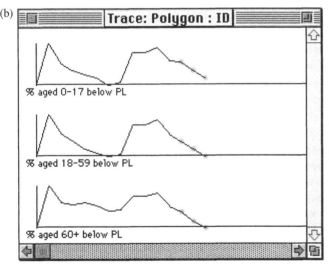

(b)

(b) The plots show that the local averages of percentages below poverty level follow similar patterns along the border. The only difference is that the levels in the 60+ group do not fall relatively as much in the central section (map interrogation shows this to be Cincinnati and counties to the west of it).

As each 'series' is plotted to its own scale, values have to be interrogated to check the implied results. Not only the last point is highlighted but the previous two as well, because the neighbourhoods of counties on which these are calculated overlap with the final selected set.

Figure 4.5 Tracing along the Southern border of the Mid-West.

Ideally, it would be good to have graphical ways of defining neighbourhoods for local statistics to explore the sensitivity of results to different definitions. Unwin (1996) shows how this may be done in a limited way using distance between centres for defining neighbourhoods. In Jason Dykes' cdv there is a prototype of a related feature called the locality probe (Dykes, 1996). In the SAGE software being developed in Sheffield by

Haining, Wise and Ma, there are special tools for aggregating areas. In MANET we have taken another approach which we call 'tracing' or 'generalized brushing'. Rather than calculating local statistics for variables across the whole region, we brush the area and calculate pre-specified statistics for selected variables as the brush is moved. (Hence the term 'generalised brushing', as far more than simple linking is involved.) The main advantage is, typically for MANET, the emphasis on a multivariate approach. Tracing produces a set of time-series-like plots in parallel, showing how the statistic for each variable selected changes as the brush is moved across the area. This is very much an exploratory technique, as the size and shape of the brush, the speed of brushing, the path of the brush, the variables selected and the statistics specified all influence the outcome. The resulting display highlights patterns and changes along the brush path (which is recorded on the map window).

Currently only running means, standard deviations, minima and maxima are programmed, but the intention is to make the system open so that any user-defined statistic may be input (such as a weighted average or LISA or G^*). In principle, any set of statistics could be calculated on the fly, providing there is sufficient computing power. A strength of the data structure used is that a change of statistic is automatically reflected in the graphic display, so that it is not necessary to rebrush the path. As linking is fully maintained between the geographic objects in the map and the statistics in the graphics, it is easy to locate the areas in the map picked out by the statistics, or, alternatively, to identify which sections of the statistics plots are influenced by particular areas.

Tracing is most valuable for brushing along naturally arising paths, such as along borders or along rivers. It was designed for working with point data and calculating statistics for all points within the brush. For regional data we have adopted the policy of including any region which is even partly in the brush. This is not an ideal solution and a more flexible one would be better. It is yet another aspect of the problem of defining local neighbourhoods.

4.3 CONCLUSIONS

Analytic statistical methods will always have problems with spatial data. Whether it is because the classical assumption of independence does not hold, or because there is such a complex variety of possible spatial interactions, other approaches are essential. Interactive graphics methods are a very attractive additional option, especially when, as in MANET, new tools are introduced, designed specifically with properties relevant to spatial data in mind. This paper has described how interactive shading for choropleth maps and weighted histograms have been implemented in MANET, as well as outlining how tools in MANET may be useful in carrying out local rather than global analyses. The value of these implementations is increased by their being part of a consistent, integrated software package with a broad range of interactive functions.

Future work will pursue a number of directions, all concerned with improving the tools and the intuitive access to them. There needs to be a better visualization and control of the definition of neighbourhoods, a yet more flexible shading interaction, an extended palette of tracing options and, as always, a more effective interface.

Acknowledgements

Thanks are due to the Deutsche Akademischer Austausch Dienst (DAAD) and the British Council for supporting the collaborative work between Augsburg, Birkbeck and Leicester.

Data

The Mid-West data set may be found at http://www.geog.le.ac.uk/argus/esdalisa. It was made available for participants at the workshop on ESDA with LISA (Exploratory Spatial Data Analysis with Local Indicators of Spatial Assocation) held in Leicester in July 1996.

Software

MANET is in development at the department of Computer-oriented Statistics and Data Analysis at the University of Augsburg. It is based on ideas from several members of the group, but the implementation is primarily due to Heike Hofmann. MANET is currently only available for Macintosh computers. For further information, look at our webpages (http://www1.math.uni-augsburg.de) and contact the first author for details of availability.

References

ANSELIN, L. (1995) Local Indicators of spatial association – LISA, *Geographical Analysis*, **27**, 93–115.

BECKER, R.A., CLEVELAND, W.S. and WILKS, A.R. (1987) Dynamic Graphics for Data Analysis, *Statistical Science*, **2**, 355–395.

DYKES, J.A. (1996) Dynamic maps for spatial science: A unified approach to cartographic visualization, in D. Parker (Ed.), *Innovations in GIS 3*, 177–187. London: Taylor & Francis.

ORD, K. and GETIS, A. (1995) Local spatial autocorrelation statistics: distributional issues and an application, *Geographical Analysis*, **27**, 286–306.

TUKEY, J.W. (1977) *Exploratory Data Analysis*, London: Addison-Wesley.

UNWIN, A.R., WILLS, G. and HASLETT, J. (1990) REGARD – Graphical Analysis of Regional Data, in *Proceedings of the 1990 ASA Statistical Graphics Section*, 36–41.

UNWIN, A.R. (1994) REGARDing Geographic Data, in P. Dirschedl and R. Ostermann (Eds), *Computational Statistics*, 315–326, Heidelberg: Physica.

UNWIN, A.R. (1995) Statistics for the Birds – Exploring Ornithological Data Sets, in M. Meyer and J.L. Rosenberger (Eds), *Computing Science and Statistics, Proc. 27th Symp. on the Interface*, 27 (85–89), Pittsburgh: Interface Foundation.

UNWIN, A.R. (1996) Exploratory spatial analysis and local statistics, *Computational Statistics*, **11**, 387–400.

UNWIN, A.R., SLOAN, B. and WILLS, G. (1992) Interactive Graphical Methods for Trade Flows, in *New Techniques and Technologies for Statistics*, 295–303, Bonn: Office for Official Publications of the EC.

UNWIN, A.R., HAWKINS, G., HOFMANN, H. and SIEGL, B. (1996) Interactive Graphics for Data Sets with Missing Values – MANET, *Journal of Computational and Graphical Statistics*, **5** (2), 113–122.

VELLEMAN, P. (1994) Data Desk, Data Description Ithaca, New York.

WILHELM, A. and STECK, R. (1997) Exploring spatial data using interactive graphics and local statistics, submitted for publication.

WILLS, G., UNWIN, A.R. and HASLETT, J. (1991) Spatial Interactive Graphics Applied to Irish Socio-economic Data, in *Proceedings of the ASA Statistical Graphics Section*, 37–41, Atlanta, GA: ASA.

WILLS, G. (1992) *Spatial Data: Exploration and Modelling via Distance-Based and Interactive Graphics Methods*, PhD Thesis, Trinity College, Dublin.

PART TWO

Virtual GIS

CHAPTER FIVE

The geolibrary

MICHAEL F. GOODCHILD

5.1 INTRODUCTION

While a geographic data set is defined as a representation of some part of the Earth's surface, many other types of information also refer to specific places on the Earth's surface, and yet are not normally included in discussions of geographic databases. They include reports about the environmental status of regions, photographs of landscapes, guidebooks to major cities, municipal plans, and even sounds and pieces of music. All of these are examples of information that is geographically referenced, or georeferenced for short, because it has some form of geographic footprint. Clearly geographic information is by definition a subset of georeferenced information.

Footprints can be precise, when they refer to areas with precise boundaries, or they can be fuzzy, when the limits of the area are unclear. For example, a municipal plan for a city has a precise footprint, but tourist information on 'Northern California' does not, because 'Northern California' is not precisely defined on the ground. Data sets can have more than one footprint, or a footprint that is more complex than a single polygon. For example, the footprint of Handel's Messiah might include the place where it was composed (London), as well as the place where it was first performed (Dublin), and perhaps also geographic references to the life of its composer.

Georeferenced information is likely to be more pertinent in some places than others. Generally, interest depends on how far one is from the footprint, and also on the size of the footprint. In California, interest in a guidebook to France will be much greater than in a municipal plan of the city of Rouen. A bookstore or public library in California might carry such a guidebook, but almost certainly no bookstore or public library anywhere in the state would have the municipal plan of Rouen. In this sense there is a sharp distinction between georeferenced data sets and other types of information: a mathematical encyclopaedia, for example, is presumably of uniform interest everywhere on the planet.

A geolibrary is a library filled with georeferenced information. Information is found and retrieved by matching the area for which information is needed with the footprints of items in the library, and by matching other requirements, although the footprints always provide the primary basis of search. A geolibrary can handle queries like 'what information do you have about this neighbourhood?', 'do you have a guidebook covering this area?', 'can I find any further information about the area in which the Brontë sisters lived?', or 'what photographs do you have of this area?' In all of these queries the geographic footprint provides the primary basis of search.

Recent years have brought an explosion of available information, notably in the form of information accessible through the World Wide Web (the 'Web'). But while certain tools exist for browsing and searching this immense information resource, more generally there is an acute shortage of methods for organizing, managing, cataloguing, and integrating data. The traditional library has perfected the cataloguing of books into a fine art, based on author, title, and concepts of subject. But the current generation of search engines, exemplified by Alta Vista or Yahoo, are limited to the detection and indexing of key words in text. By offering a new paradigm for search, based on geographic location, the geolibrary might provide a powerful new way of retrieving information.

Consider an example of crisis management. The Oklahoma City bombing of April 1996 created an immediate need for information about the construction of the building, nearby hospital resources, evacuation routes, the locations of underground gas pipes that might have been damaged, the potential for adverse weather to affect the rescue effort, the identities of employees, and much more. The common element in all of this information was the geographic location of the explosion – if this could have formed the basis of search, the relevant information might have been assembled much more rapidly.

The purpose of this paper is to explore the concept of the geolibrary, and its relationship to the traditional concerns of GIS research, as a thought-experiment. Many elements of the geolibrary are present in traditional institutions, data centres, and earlier research efforts, and these are reviewed in the next section of the paper, which explores the degree to which geolibraries can be said to exist. This is followed by a further discussion of the value and impact of a functioning geolibrary. The subsequent section identifies the components of the geolibrary, and this is followed by a discussion of the important research issues that must be solved if the geolibrary is to be achievable.

5.2 DO GEOLIBRARIES EXIST?

A user of a geolibrary might begin the process of search and retrieval by pointing to a map or a globe. Physically, this would mean having a large map or globe in the library's entrance; but it would be impossible to link the action of pointing to a map or globe with the card records of appropriate items in the library. It would be impossible also to provide for many different levels of detail, in order for the library user to be able to specify areas as large as France or as small as a single city neighbourhood.

An alternative approach would be for the user to specify a placename, and to provide a fourth kind of indexed card catalogue, in addition to the conventional indexing by author, title, and subject. But placenames only work to a degree. There would be no way, for example, to link entries under France with entries for the various French Departments, or the names of European regions that include part or all of France. Geographic placenames do not lend themselves readily to use as subject headings in card catalogues, because of the complex set of hierarchical and other relationships that exist between geographic locations.

Essentially, it is impossible to build a physical geolibrary, although conventional map libraries come as close as it is possible to come. In a digital world, however, these problems disappear. The user of a digital geolibrary can be presented with a globe; can zoom to the appropriate level of detail; can access lists of placenames and see their footprints; and can move up or down the placename hierarchy using links between places.

Moreover, a digital geolibrary solves the problem of physical access, if the services of the library are provided over a universal network like the Internet. In a digital world, the differences between storage media, which have made it difficult for the conventional physical library to handle music, maps, or photographs, disappear. And, finally, the collection of a geolibrary can be dispersed – a digital geolibrary can consist of a collection of servers, each specializing in materials about its local region.

5.3 THE ALEXANDRIA DIGITAL LIBRARY

In the past few years there has been an increasing awareness of the technical feasibility of digital libraries – that is, libraries that emulate the services of conventional physical libraries, but using digital technology, with access through digital communication networks. Such digital libraries would be of immense benefit in extending access to library services to anyone connected to digital networks; would add the capability to search information by content; and would provide information in digital form, allowing users to analyze it further, using suitable application software.

The Internet has already moved us closer to the vision of universal access via electronic networks (although many would argue that it has simultaneously widened the gap between have's and have-nots). A simple calculation shows that the textual contents of a typical, large research library, when digitized in a simple code such as ASCII, amount to order 10^{13} or 10^{14} bytes, and thus lie within today's storage capacities. In short, a digital library is sufficiently feasible to justify further research. In the USA in 1994, three federal agencies (the National Science Foundation, NASA, and the Advanced Research Projects Agency) jointly initiated a Digital Library Initiative, funding six projects for a period of four years. The Alexandria Digital Library (ADL), based at the University of California, Santa Barbara, is one of those projects.

The primary goal of ADL is to develop the services of a distributed library of georeferenced materials. Smith *et al.* (1996) describe ADL, and further information and descriptive reports, as well as access to the current prototype, are available via http://alexandria.ucsb.edu.

In the first phase of the ADL project, from October 1994 to March 1995, we developed a Rapid Prototype using existing off-the-shelf software, including ArcView 2, Tcl/Tk, and Sybase. This was a stand-alone prototype, distributed on CD for Windows. Our first Web prototype was completed in 1996, using a standard Web browser but providing full Internet access. ADL is currently (April 1997) working on a new Web prototype using Java and VRML tools.

At this point in its development, ADL satisfies some of the requirements of a geolibrary, but by no means all. A subsequent section, which reviews the necessary components of a geolibrary, includes an evaluation of the relevant components of ADL in its current prototype.

5.4 OTHER PROJECTS

ADL is by no means unique in its attempt to provide the services of a geolibrary. The DLI project at the University of California, Berkeley, also focuses on multimedia and georeferenced materials, although it has additional emphases (information on all of the DLI projects can be accessed through the ADL web site). Many developers of software for processing maps, images, and similar materials have recognized the need for additional modules to manage what is often a large and rapidly growing resource of data sets. ESRI, for example, offered MapLibrarian as a component of early versions of its popular ARC/ INFO software, and Intergraph provides its own document management software to meet the needs of its larger CAD and GIS customers. Elements of the geolibrary concept are present in many systems for management, browse, and dissemination of geographic data, including EOSDIS (the Earth Observing System Data and Information System), GC-DIS (the Global Change Data and Information System), the FGDC's (Federal Geographic Data Committee's) National Geospatial Data Clearinghouse, and many others. One of the first such projects was MARCMAP (Morris *et al.*, 1986), at the University of Edinburgh beginning in the early 1980s (and see http://www.geo.ed.ac.uk/~dcf/gisa/Cartonet.html); and see also Goodchild and Donkin (1967).

5.5 THE VALUE OF A GEOLIBRARY

In the first instance, a geolibrary could provide an interesting new access mechanism to the contents of a conventional library – for example, a tourist planning to visit Paris would have a powerful new way of finding guidebooks. But the full power of the geolibrary lies in its ability to provide access to types of information not normally found in libraries.

First, the geolibrary would be a multimedia store. Because everything in a digital library is stored using the same digital medium, there are none of the physical problems that traditional libraries have had to face in handling special items like photographs, sound, or video.

Secondly, the geolibrary could focus on building a collection of items of special or local interest. In a networked world there is no need for libraries to duplicate each other's contents – it may be sufficient for one digital copy of the Rouen municipal plan to be available in a digital geolibrary located in Rouen, or perhaps in Paris, provided other geolibraries know it is there. When material is georeferenced, there is a clear advantage to locating its digital version on a server close to, or within, the geographic footprint of the material, since that is where interest in it is likely to be highest.

Thirdly, because interest in it tends to be geographically defined, much georeferenced information is unpublished or fugitive. The geolibrary provides an effective mechanism for collecting such information in special, local collections, and making it widely available.

In summary, the contents of a geolibrary would be very different from those of a conventional physical library. They would be dominated by multimedia information of local interest – in fact precisely the kinds of information needed by an informed citizenry, one that is deeply involved in issues affecting its neighbourhood, region, and planet. Because its contents would be different, a geolibrary might attract an entirely new type of library user. At the same time, geolibrary technology would be of great benefit to major utilities, resource industries, and other private sector corporations with large information management problems and a strong geographic context. Such corporations, running geolibraries over intranets, might provide the market needed to stimulate commercial development of geolibrary technology.

5.6 THE COMPONENTS OF A GEOLIBRARY

While there are sharp differences in approach and scope between the various projects that have attempted to construct versions of a geolibrary, there is now among them a degree of consensus on its basic nature.

5.6.1   The browser

Browsers are specialized software applications running locally in the user's computer, and providing access to the Web. The Alexandria Digital Library project's current prototype is an example of a geolibrary using standard Web browser software, such as Microsoft's Internet Explorer. Ideally, a geolibrary browser would be more specialized, but universally available. It would have 3D extensions (e.g. a VRML plug-in) to allow the user to interact with the surface of the globe, and for specialized interactions such as 'fly-bys'. To avoid having to communicate large amounts of repetitive data, it might use a 'hybrid' approach by storing the basemap and gazetteer locally (see below). Finally, like current Web browsers it would provide a uniform 'look and feel' across a wide range of computer hardware.

5.6.2   The basemap

The basemap provides the image of the Earth on which the browser's user can specify areas of interest, and on which footprints of library items can be displayed. Its level of geographic detail defines the most localized search possible. It should include all the features likely to be relevant to a user wanting to find and define a search area, including major topographic details and placenames. The importance of such features will vary between users, as will levels of detail, so it will be necessary to establish protocols that allow use of specialized basemaps for particular purposes.

Basemap information may be voluminous and is not likely to change frequently, so rather than transmit it repeatedly from a server it may be more efficient to store it locally, in a specialized 'hybrid' browser.

Suitable basemaps include digital topographic maps, and also images of the Earth's surface. Additional detail can be provided by digital elevation data, so that the basemap provides a close resemblance to the actual surface of the Earth. The availability of such data is discussed below.

5.6.3   The gazetteer

'Gazetteer' is a technical term for the index that links placenames to a map. Gazetteers are commonly found in the backs of published atlases and on city maps. In a geolibrary the gazetteer allows a user to define a search area using a placename, instead of by finding the area on the basemap. The gazetteer may include placenames that are not well defined.

For use in a geolibrary a gazetteer must include extents, or digital representations of each placename's physical boundary. Links between placenames allow searches to be expanded or narrowed – they can be vertical, identifying places that include or are included by other places, and also horizontal, identifying neighbouring places.

A gazetteer protocol would allow specialized gazetteers to be used in particular applications. Like basemaps, gazetteers are not likely to change rapidly, so could be stored locally in 'hybrid' browsers. The availability of suitable data is discussed below.

5.6.4   Local collections

A geolibrary could operate in a standalone form, with the browser providing access to a collection of items of information stored on the same machine, or on a machine linked logically to it. The browser would display the availability of information in the form of icons superimposed on the basemap, with associated details, and the user would be able to retrieve and manipulate the items.

5.6.5   Server catalogues

The contents of a specialized geolibrary collection could also be made available to browsers over a network. Each collection would be maintained by a server. A user would initiate a search with a browser, and transmit a specification of the search to a server, which would then respond with an indication of the items in its collection satisfying the search. These would then appear as symbolic icons superimposed on the browser's basemap, and as items in a list with additional details. The user could request more detail, including the downloading of the item itself, by selecting an icon from the basemap or from the list. In addition, a server might provide a generalized 'browse' version of the item to avoid forcing the user to wait for a lengthy download (see below).

For this mechanism to work, the user must give search specifications in a standard format, following an agreed protocol. The basis of such protocols already exists in standards such as MARC and the Federal Geographic Data Committee's Content Standards for Digital Geospatial Metadata (http://www.fgdc.gov), and in projects such as ADL.

5.6.6   Basemap data

Several suitable sources of data exist for basemaps:

■ *Digital topographic data.* Available for the entire land area of the planet at 1:1 000 000 in the Digital Chart of the World (in very rough terms, a map at a given scale shows features that are at least 0.5 mm across at the scale of the map; this means that a 1:1 000 000 map would show features at least 500 m across). Data are also available for smaller areas at larger scales – e.g. for the continental U.S. at 1:100 000, or for the UK at 1:10 000.

■ *Imagery.* Available from the Landsat satellite at 30 m resolution; from the SPOT satellite at 10 m resolution; from Russian satellites; and in late 1997 for selected areas at 1 m resolution.

■ *Digital elevation data.* Available for parts of the USA at 30 m resolution; for the entire planet (classified) at approximately 100 m resolution; for the entire planet (public domain) at 10 km resolution; expected to be available in late 1997 for the entire planet at 30 m resolution.

5.6.7   Gazetteer data

A combination of public sources (the Geographic Names Information System of the USGS, and the Board on Geographic Names) can provide about 7 million point records. There are no significant sources of information on feature extents. There are significant problems to be overcome in dealing with varied alphabets and languages.

5.6.8 Search across geolibraries

On the Web one has access to numerous search engines to assist in the process of identifying the server most likely to contain a given set of information. Search engines such as AltaVista work by 'crawling' the Web looking for servers, and extracting appropriate key words that can be indexed. This same mechanism would work for placenames, but would be subject to the same problems identified earlier that limit the ability of a placename index to find georeferenced information successfully.

Instead, a geolibrary must use alternative mechanisms to identify the servers to be searched. One would be the creation of a union catalogue – the US National Geospatial Data Clearinghouse is a current example – that contains a record for each item available in any server whose contents have been registered with it.

A second mechanism would rely on the development of a set of rules for predicting the server(s) most likely to contain a given item. Such rules could depend on geographic location – the server most likely to contain an item about location X would be the one closest to X, for example. This approach will work only if the set of servers is small, and well organized.

Thirdly, the search could visit every known geolibrary for every search. This is clearly time-consuming, and fails to 'scale' as the number of geolibraries increases.

Fourthly, one could introduce extensions to the mechanisms currently used by the Web crawler agents acting for the search engines. These would include explicit identification of placenames, and mechanisms for identifying the co-ordinates of footprints embedded in text. This option seems the most powerful, but would require development of a new generation of crawlers and associated protocols.

5.6.9 Browse images

The bandwidth available between a server and a client varies enormously, depending on many factors. Geographic information can be voluminous, and it is easy for a single multiband image of a small part of the Earth's surface to exceed a gigabyte. Thus, a geolibrary operating over an open network will have to consider the problems of limited bandwidth in its design. At the very least, a user should be warned about the probable time required to download an information object.

One general solution to these issues is to support progressive transmission. In such schemes, a coarse and thus comparatively small version of the information is sent first, followed by increasing levels of detail. In the case of an image, the technology of wavelet decomposition provides a suitable mechanism. In ADL, we also use a concept of a browse image or thumbnail sketch, a small raster containing a very generalized version of the image, and make this available to the user who expresses interest in a particular information object. There are problems, however, in generalizing the idea to types of information other than images.

5.6.10 Objects and wrappers

In the conventional library, information is packaged in convenient bundles as bound volumes. In the geolibrary, each information object, or 'bag of bits', must be packaged inside an appropriate digital wrapper, which provides the information needed by the system to ensure appropriate handling. The wrapper must identify, for example, the internal format of the data in sufficient detail to allow the system to open and display its contents. ADL currently supports only a few formats. On the other hand, it seems very unlikely, given the philosophy of the geolibrary and other Web-based systems, that one would ever be able to impose or achieve uniform format standards. In the long run, therefore, a geolibrary must be able to support a large number of alternative information object formats.

5.7   RESEARCH ISSUES

Besides the technical problems identified above, a large number of research questions need to be addressed.

1   What are the legal, ethical, and political issues involved in creating geolibraries? What problems must be addressed in the area of intellectual property rights, and what are their technical implications?

2   What are the economics of geolibraries? Who should pay for their creation and maintenance? What elements might be free (funded by the public sector, or by the private sector as loss-leaders)? Where are the potential revenue sources?

3   What is the appropriate scale for a geolibrary? How much material should be assembled on one server, and what geographic area should it cover? What are possible transition trajectories between current arrangements and a future distributed geolibrary?

4   What institutional structures would be needed by a geolibrary? What organizations might take a lead in its development? What is a possible timetable?

5   What models exist for the creation of geolibrary metadata? What are the advantages of following a library cataloguing model, versus a fully voluntary model where metadata is provided by the custodian? How much metadata is needed?

6   What techniques can be used to locate the geolibrary most likely to contain a certain item (see 5.6.8 above)? Research would include experiments with working prototypes of the most promising approaches.

7   What are the cognitive problems associated with using geolibraries? Is it possible to construct a geolibrary that is useful to a child in Grade 3, for example? What protocols would users have to master, and what problems would occur in using geolibraries across cultural or linguistic barriers?

8   Is it possible to compare the geolibrary with conventional libraries, both physical and digital? What are appropriate metrics of comparison, and what general principles can be learned from their application?

9   What protocols are needed to support the geolibrary? What role should existing institutions play in developing them? Are existing protocols such as HTTP and VRML sufficient? What protocols are needed for metadata, base maps, and gazetteers?

10   Do the data sets necessary to support the geolibrary exist? What initiatives are needed to develop or compile them?

11   Is it possible to develop a new generation of search engines and agents that can scan the Web for georeferenced materials? What protocols would be needed to make such agents effective?

12 Can the concept of a footprint be extended to the indeterminate or fuzzy case? What mechanisms are needed to define, store, and use fuzzy footprints?

13 Is a Boolean approach to the intersection of query and information object footprints appropriate, or is it possible to develop metrics of fit between them, and to rank information objects on that basis?

14 Can the concepts of browse images and progressive transmission be extended from the raster case, notably to vector data sets, and non-geographic information objects? What other options exist for intelligent use of bandwidth in geolibrary applications?

5.8 CONCLUSION

Projects like ADL, and the concept of a geolibrary, are bridges between the GIS and library communities. Libraries have centuries of expertise in such functions as quality assurance in building collections, and abstracting of information to support search, that are as relevant for geographic information as they are for any other type. Yet, as in so many other areas, traditional libraries have been constrained by physical limitations to focus on certain types of information that are comparatively easy to handle. The digital revolution has changed all that, and has made geographic information potentially as accessible and easy to handle as other types. At the same time digital communication networks have provided the potential for universal access to information. Thus, we find ourselves in the GIS research community at the beginning of a period of exciting collaboration with the library and information science communities.

At this point in history, libraries are faced with apparently insurmountable problems (Hawkins, 1996). The published corpus of humanity is growing rapidly, and doubling in not much more than ten years. Journal prices continue to rise at well above the rate of inflation. Library budgets are contracting, and libraries are faced with unprecedented problems of security. The pressures to find new approaches, and to take advantage of new technologies, are high.

I would like to close with four points. First, the concept of georeferenced information extends the domain of digital geographic information well beyond more conventional digital maps and images, to include a vast array of information with definable geographic footprints. This new class offers an exciting new domain for the GIS research community.

Secondly, the geographic key is a powerful way to organize and integrate information, whether it be geographic or georeferenced. It has been difficult to exploit this particular key in the past because it is based in a continuum, whereas the more successful keys of information management – author, title, subject – are inherently discrete. But in a digital environment this problem is much less severe. Moreover, the Web has shown how important it is to develop new ways of integrating information, to support effective search over a rapidly expanding but inherently unorganized information base.

Thirdly, geolibrary projects like ADL demonstrate the potential for radically different approaches to information management. Information access mechanisms ultimately determine the kinds of material that can be served by systems such as libraries – new mechanisms thus offer the potential for storing novel types of information, and for reaching new user communities.

Finally, I hope this paper has identified a number of challenging research issues. Although progress has been made, we are still a long way from being able to build a functioning geolibrary, for reasons that are both technical and institutional. The GIS research community has much to offer in both areas.

References

GOODCHILD, M.F. and K.M. DONKIN (1967) A computerized approach to increased map library utility, *The Cartographer*, **4**.

HAWKINS, B.L. (1996) The unsustainability of the traditional library and the threat to higher education. Paper presented at *the Stanford Forum for Higher Education Futures*, The Aspen Institute, Oct. 18.

MORRIS, B.A., BARTLETT, D.J., DOWERS, S., GITTINGS, B.M., HEALEY, R.G., IRVINE, J.J.A. and WAUGH T.C. (1986) Cartographic information retrieval and the creation of graphic indices from database records. Paper presented at the *British Library Research and Development Department Symposium*, Cranfield Institute of Technology, 27 July, Department of Geography, University of Edinburgh.

SMITH T.R., ANDRESEN, D., CARVER, L., DOLIN, R., GOODCHILD, M.F. and others (1996) A digital library for geographically referenced materials, *Computer*, **29** (5), 54+ (corrected author list appears in **29** (7), 14).

CHAPTER SIX

The KINDS project: providing effective tools for spatial data accessibility and usability over WWW

ADRIAN MOSS, JIM R. PETCH, A. HEPTINSTALL, K. COLE, C.S. LI,
K. KITMITTO, M.N. ISLAM, Y.J. YIP AND A. BASDEN

6.1 INTRODUCTION

The mere existence of rich data resources does not guarantee that they are accessed and used effectively. Effective use is dependent upon awareness, accessibility and usability of data resources. Also, to be able to use a data set users must be appropriately trained. Many spatial data sets are large and complex, requiring expert skills and knowledge. Raper and Green (1992) attribute an acute shortage of spatial data handling skills to a lack of awareness of the potential usefulness of spatial data, the need to learn concepts associated with the new technology and to develop specific technical skills. The skills shortage is compounded because existing skills are not readily transferable and retraining may take 18 months to three years (Davies and Medyckyj-Scott, 1994).

These factors can restrict data-use to a relatively limited number of technically highly-experienced users. The problem is exacerbated by potential users who, lacking the necessary skills, may even be unable to evaluate properly the usefulness of a spatial data set. This can create a self-perpetuating cycle which restricts the number of spatial data users to those already experienced. Potential new users may be reluctant to invest time in learning how to use spatial data handling tools before ascertaining whether the available data sets suit their needs.

The Knowledge-based Interface to National Data Sets (KINDS) Project is a response to these problems. Its aim is to reduce the overhead of using spatial data sets by tackling three key problems: low *awareness* of the content, coverage and structure of the data sets; poor *accessibility* of data, which is held in different formats and manipulated using different tools; and low levels of data *usability* which requires users to be equipped with the knowledge they need to use data effectively (Petch *et al.*, 1997). KINDS has developed a series of interfaces on the world wide web (WWW) to address these problems. This paper comprises three sections. In the first, the user-orientated methodology employed

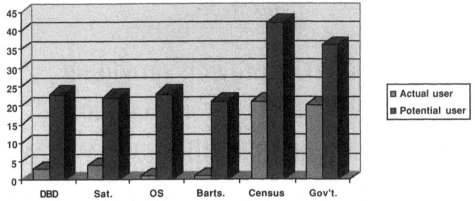

Abbrev.: DBD – digitized boundary data; Sat. – satellite imagery; OS – Ordnance Survey samples; Barts – Bartholomew's data; Census – 1991 Census; Gov't – Government Surveys: after Li et al. *(1996).*

Figure 6.1 Actual vs. potential data set use for research and teaching. Source: *Proceedings, Third International Conference/Workshop on Integrating GIS and Environmental Modelling,* by kind permission of NCGIA.

to design the system is described. In the second the development of multiple data-set search tools and inter-mediary systems which help the user work with complex data sets is outlined. Finally, an evaluation of the effectiveness of the KINDS WWW-based system is undertaken. Thirty-eight academics are surveyed and the ease of use of the system is assessed.

6.2 KINDS AND MIDAS

The academic community has invested heavily in spatial data sets. These include the 1991 Census and associated map data, Bartholomew's digital maps and satellite images from the Landsat and SPOT programmes. To avoid duplication and to centralize support and hardware resources, these data sets are held by strategic data libraries, including Manchester Information Data sets and Associated Services (MIDAS).

Although some data sets, such as the Census, attract large numbers of users, their uptake and that of others in relation to the numbers of potential users has been only partial. Figure 6.1, based on a survey of 78 academics, shows a considerable potential. The focus of higher education is shifting, encouraging research and increasing the numbers of students catered for by resource- and distance-based teaching. A more marked increase in the demand should, therefore, be expected. The KINDS Project focuses on increasing the numbers of users of the data sets held by MIDAS. The wider set of issues associated with improving the ease of use of spatial data sets is of importance to the development of user interfaces to complex spatial systems and WWW-based services in general.

6.3 PROBLEM ANALYSIS – IDENTIFYING THE REQUIREMENTS OF USERS

The analysis of large, complex data sets using GIS is a knowledge-intensive activity requiring extensive skills and training. The key to increasing their ease of use lies not

Table 6.1 User analysis methodology.

Technique	Method	Purpose
General user survey	Open ended questionnaire interviews	Establish the range, ability and subject areas of interest to academics
Computer use survey	Multiple choice postal questionnaire	Establish potential users' expertise with computers and preferred user interfaces
Expert knowledge elicitation from a) spatial data experts b) MIDAS support staff	Task analysis interviews	a) Establish expert task models for handling spatial data b) Identify common user problems and their solutions

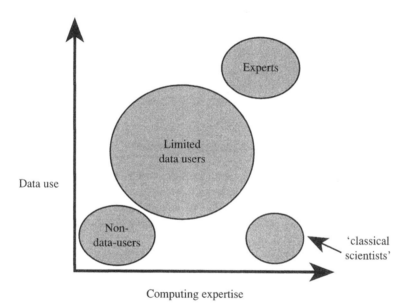

Figure 6.2 A conceptual distribution of user groups.

only in a thorough understanding of the data set, but also in understanding the potential user (McGraw, 1992). Three information-elicitation exercises were undertaken prior to the systems development (Table 6.1). An initial interview survey asked respondents from selected disciplines about their teaching and research interests and use of spatial data sets. Four types of user were identified from this survey by classifying the extent of their data set use and technical expertise (Figure 6.2).

The *four* groups of user may be described as:

1 *Non-data-users*. By far the largest number of respondents, characterized by low levels both of computing skills and data set use.

2 *Limited users*. The second most numerous group composed largely of social scientists who had exploited some (often survey) data sets. Many members had a particular interest in mapping the results of survey analysis.

3 *Experts*. Technically highly-competent and experienced data users, these users based much of their research and teaching around the use of large spatial data sets.

4 *Classical scientists*. Working in fields such as physics, chemistry or engineering. Respondents tended to be familiar with computer applications but had little interest in spatial data sets.

Figure 6.2 shows a conceptual distribution of these users, based upon their use of spatial data sets and computing knowledge. Non-data-users and limited users account for the largest numbers of potential users. Tradtional scientists perceived little need for spatial data and experts were judged already well equipped. The most efficient method of increasing the number of users is, therefore, to introduce the 'non users' to data sets and increase the use of the limited users. The development of the KINDS system has been targeted towards the requirements of these potential users. The approach to increasing the use of spatial data amongst the target groups has been two-pronged. First, there has been an effort to increase the accessibility of spatial data sets. This has been accomplished by constructing search tools and interfaces allowing users easily to identify and manipulate the data sets. Secondly, it is also necessary to introduce potential users to data, bodies of theory and examples of application and methodologies. Accordingly, a programme of knowledge-elicitation interviews with experts was begun to distil the skills and background information required. There has also been an effort to identify and resolve common problems faced by users. Knowledge-based help systems which will encapsulate this data within a WWW interface are presently under development.

6.4 THE KINDS SYSTEM

The systems developed and tested to date are designed to tackle the problems of low awareness and poor accessibility of data. Low awareness is tackled by search engines designed to allow the user to identify easily the contents and structure of data sets. Poor accessibility is countered by providing high level tools with which the user can manipulate the data sets.

Data sets held by MIDAS have been developed by different organizations often over considerable periods of time. This has resulted in deployment of data sets to the user community via many poorly integrated routes. For example the *de facto* standard software for extracting 1991 Census data is SASPAC, a package originally developed to manipulate the 1981 census (LGMB and LRC, 1992). Other data sets are available through proprietary software or stored as raw data files. New users must, therefore, learn how to operate many unfamiliar software packages and at the same time must also identify the contents, limitations, structure and suitability of the data sets themselves. The key to increasing the numbers of users lies in amalgamating access to data sets with knowledge about their use. The KINDS system unifies access to data sets behind a common WWW user interface. At the time of writing, the system provides access to Bartholomew's digital mapping, the 1991 Census and associated mapping and SPOT satellite images.

6.4.1 Integrated search engines

To enable the user to identify available data, two search tools have been developed (Li *et al.*, 1996; Moss *et al.*, 1997). A spatial search engine allows the user to select an area of interest by clicking on a map. The location upon which the user clicks is converted into six figure national grid co-ordinates. These co-ordinates are compared with Bartholomew's and Census map data sets and 'footprints' of satellite images. Data falling within the proximity of the search location is listed. The user may examine 'quick looks' of the various data sets. Quick looks are designed to show the content and scope of the data set.

A free text search engine may also be used to search the data sets. By typing place names or data categories (for example 'age' or 'motorway') the search engine will again return quick looks of data.

6.4.2 Intermediary systems

High-level data access tools have been implemented by developing intermediary systems between the user and data-manipulation packages. The intermediary is analogous to a specialist librarian helping an inexperienced reader with a complex bibliographic search (Li, *Unpublished manuscript*). A librarian can direct a reader to relevant books, help search the on-line catalogue or request items from another library. The reader need not familiarize themselves with each low-level task, but instead concentrate on the high level task of locating information. Intermediary systems have been built to allow the user to map the complex Bartholomew's and census data sets. The user is only required to make high-level decisions about the area and subjects they wish to map, while the intermediary handles the tasks of activating the mapping package, loading it with the required data and the specifics of the cartographic output. The result is displayed to the user over the WWW (Figure 6.3).

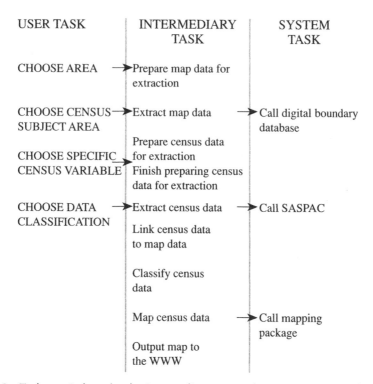

Figure 6.3 Tasks carried out by the intermediary system for mapping census data.

6.5 INTERFACE EVALUATION

6.5.1 Research design

To test the effectiveness of the search tools and mapping systems an evaluation survey has been carried out. The survey tested the ease of use of the system by timing how long it took a user to carry out a defined set of tasks using it. Respondents were asked to use the search tools to find all the data held on the system about a given area and then make two maps of that area. One map was made using the Bartholomew's map maker, and one with the census map maker. The 38 respondents were academics and students classified into three of the groupings shown in Figure 6.2 (non-data-users, limited data users and experts). The length of time taken to perform each stage of the task was noted, as were any problems or queries raised by the respondents.

6.5.2 Method

The WWW browser used for the test was Internet Explorer 3.0 (Microsoft Corporation, 1996). The majority of tests were carried out on a P200 MMX PC running Windows 95 and a small number were carried out on a P200 PC running Windows NT Server 4.0. For the purposes of this evaluation the user interface is the same upon both systems.

Each respondent was read a prepared text of instructions and timed during the test. Timing began when a Web page loaded. The respondent read the page and considered what to do next. When they had made their decision they told the examiner, who stopped timing and noted the result along with any other comment. The halting of timing in this way is necessary to ensure that the test logs the length of time taken by the user and not by the system or network.

The respondent first searched for all information held in the KINDS system about one of four predetermined places (Swansea, Gloucester, Hereford or Chester). The respondent could use either the free text or map-based search tool. Secondly, the user made two maps of the target area using the Bartholomew's and Census mapping interfaces.

6.5.3 Results

Most users chose the free-text search engine, although a higher proportion of expert spatial data users chose the map-based search. However, the search process proved longer when using the map-based tool (mean average time using the free-text search engine was 24.8 seconds, while the map-based search took 39.7 seconds, on average). The main reason for this was that some users found it hard to locate their targets using the map. Furthermore, of the eleven respondents who initially chose the map-based search, six subsequently abandoned it and chose the free-text search.

Overall the KINDS system provides very fast access to data sets, the average total time taken varies from 135.2 seconds, (non-data-users) to 86.7 seconds (experts).

Table 6.2 indicates the average time taken to complete each stage of the test. Figure 6.4 depicts the cumulative time taken by the different user groups to complete the test. Overall, 'experts' complete the test significantly faster than the other two groups. However, when timings for the individual stages are examined, the most significant difference is in the time taken to use the Bartholomew's map maker (Table 6.2).

Table 6.2 General factorial ANOVA of time taken for each individual stage.

Time stage	Type of user			Significance
	Non-data-user	Data user	Expert	
Search	33.1	27.1	25.9	0.466
Bartholomew's	49.9	46.4	23	0.000
Census	52.2	49.9	39.8	0.217
Total time	135.2	123.5	86.7	0.012

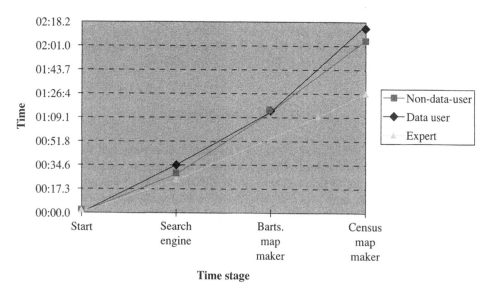

Time stage

Figure 6.4 Cumulative timings against user type.

6.5.4 Discussion

KINDS provides fast access to data. Typical access times amount to no more than a few minutes, the actual access time being largely dependent on network speeds. A user can create a complex map, including – in the case of the census – combining map and attribute data, very quickly. Even an experienced spatial data user could not perform the same process in this time. A new user would have to undertake a significant period of training before being able to carry out the same tasks.

The value of these interfaces is expressed also in the fact that non-data and data users show hardly any difference in access times. Experts show slightly faster times, mainly when using the Bartholomew's mapper.

The difficulties in these interfaces can be examined by looking at the stages which slowed the retrieval process. Two main problems areas were identified. The first difficulty was in the use of the map-based search engine. Some users were not familiar with the exact spatial location of the data they wished to retrieve. The second problematic section was the first screen of the Bartholomew's mapper series. This showed the work flow of options when using the mapping wizards. Novice users spent some time studying this, whilst experts moved on quickly. The differences in lengths of time taken to view this screen largely accounted for the difference between experts and non-experts in the length of time to use the system.

In each case there are arguments for the retention of the screens. Although users took some time to understand the flow diagram, the overall times taken to use the mapper are still low. At present research is being carried out into the construction of search tools which overcome some of the problems associated with discovering information about spatial data. An approach being explored is the development of a detailed gazetteer. The gazetteer will link the 26 000 locations in the Bartholomew's gazetteer with the spatially corresponding areas in census, national grid geographies and satellite image footprints. As a result, every gazetteer location will be linked to a spatial unit in each of the data sets supported by KINDS.

A further refinement to the test reported here would be to examine how much quicker users got when using the interfaces a second or third time, and so on. However, as the range in timings was very narrow and the difference between novice and expert users relatively small, the value in such testing is likely to be limited.

6.6 FUTURE DEVELOPMENTS

We have identified a large, unmet, demand for spatial information, and shown that the KINDS interfaces make a considerable improvement to access to data sets. The intermediary systems developed devolve some of the mundane tasks of making maps to the computer, but do not, however, increase the amount of geographical research being carried out. Simply improving access is not enough. Users also require access to other types of material. Users require access to help and didactic systems describing the technical details and scope of data sets (such as coverage, date/time of capture and data quality) and a wider set of materials about the use of data sets. This wider set of materials could include derived data sets, academic literature and teaching materials. The development of a knowledge-based help system is underway and further funding has been obtained under the Joint Information Systems Sub-Committee Technology Application Programme to develop a system which draws derived data sets and teaching materials back from the user community. This will create a growing and self-perpetuating data resource. These materials will be used to link data to bodies of theory and examples of data use.

Acknowledgements

We would like to thank all the survey respondents for their help. The Knowledge-based Interface to National Data Sets (KINDS) project is funded by the Joint Information Systems Committee (New Technology Initiative NT-107, J-TAP award 3/302) and a grant from the Information Systems Sub-Committee of the UK Higher Education Funding Councils.

Bartholomew digital data is Copyright© Bartholomew and available to UK academics from MIDAS under the terms of the CHEST license agreement. We wish to thank Tim Rideout of Bartholomew for his support. Census data is Crown Copyright© and the ED-Line Digitised Boundary Data set is Copyright© Crown, Ordnance Survey and Post Office. Both data sets are available to the academic community through MIDAS and the UKBorders Project (University of Edinburgh, Data Library).

The prototype census mapping software presently uses GD.pm a Perl Library which is © 1995 Lincoln D. Stein. GD.pm is in turn taken from GD.c distributed by www.boutell.com © 1994, 1995 Thomas Boutell and Quest Protein Database Center, Cold Spring Harbor Labs.

References

DAVIS, C. and MEDYCKYJ-SCOTT, D. (1994) GIS usability: recommendations based on the user's view, *International Journal of Geographical Information Systems*, **8** (2), 175–189.

LI, C.S. (unpublished manuscript) *Spatial Data*: Access and Usability via the Internet.

LI, C.S., BREE, D., MOSS, A. and PETCH, J. (1996) Developing Internet-based User Interfaces for Improving Spatial Data Access and Usability, in *Proceedings of the 3rd Int. Conf. on integrating GIS and environmental modelling*, 21–25 Jan., Santa Fe NM, USA., WWW and CD-ROM, http://www.ncgia.ucsb.edu/conf/SANTA_FE_CD-ROM/main.html.

LOCAL GOVERNMENT MANAGEMENT BOARD AND LONDON RESEARCH CENTRE (1992) *SASPAC User Manual*, Part 1.

McGRAW, K.L. (1992) *Designing and evaluating user interfaces for knowledge-based systems*, Chichester: Ellis Horward.

MICROSOFT CORPORATION (1996) Internet Explorer version 3.0 documentation, http://www.microsoft.com/ie.

MOSS, A., PETCH, J., LI, C.S., BREE, D., COLE, K., YIP, Y.J., JOHNSTON, A., KITMITTO, K. and BASDEN, A. (1995) Access Large and Complex Data sets via WWW, in *New Techniques and Technologies for Statistics II*, 85–96 [Amsterdam: IOS Press].

PETCH, J., MOSS, A., JOHNSTON, A., YIP, Y.J., ANDREW BASDEN, A., LI, C.S., COLE, K. and KITMITTO, K. (1997) Knowledge-based Interfaces for National Data Sets: The KINDS project, in *New Techniques and Technologies for Statistics II*, 193–202 [Amsterdam: IOS Press].

RAPER, J. and GREEN, N. (1992) Teaching the principles of GIS: lessons from the GISTutor project, *International Journal of Geographical Information Systems*, **6** (4), 279–290.

CHAPTER SEVEN

Methodologies and tools for large-scale, real-world, virtual environments: a prototype to enhance the understanding of historic Newcastle upon Tyne

SCOTT PARSLEY AND GEORGE TAYLOR

7.1 INTRODUCTION

There is a growing demand for comprehensive 3D spatial information, for improved inventory, analysis, prediction, simulation and visualization. Increasing attention is being given to the requirements of comprehensive information systems, such as Geographical Information Systems (GIS), to move from the utilization of predominantly 2D digital mapped data sets to those that are fully 3D. Most commercially available GIS still cannot handle true 3D data, but most Computer Aided Design (CAD) systems can. These CAD systems have all the required functionality for manipulating the data, but have restricted data structure (Kraak, 1989). Some recent GIS, developed by major CAD vendors, have overcome this direction problem. A CAD product by such a vendor is used in this work to construct the initial 3D model.

One area of particular interest for geomaticians is the modelling and effective acquisition of 3D data of the urban environment (Templfi, 1997). The reconstruction of historical urban monuments and their exploration, sometimes referred to as virtual heritage, is the particular urban application of 3D digital mapped data examined here. Furthermore, the visualization and manipulation of urban 3D data is commensurate with this interest. Moreover, spatial information is processed visually in the main by its user through its topological relationships (Petch, 1994). The totality and complexity of topological relationships in 3D spatial data is far richer than that of two-dimensional (2D) data. Two aspects of this demand are considered for the construction of an historic urban environment. First, a necessary and unorthodox method of data acquisition, using single photographic images, and, secondly, its subsequent manipulation and visualization using virtual reality (VR) software.

Current VR visualization applications are often used solely to generate fictitious imagery. This project presents an approach for the construction of virtual reality environments from measurements and observations of real world objects. This will provide a more accurate representation of these objects than the usual 'faithful reproductions' offered in many VR townscapes. The aim is to construct a fully three-dimensional digital townscape of Newcastle upon Tyne throughout the various stages of its history. This includes the representation of buildings in their correct historical location, the developing infrastructure of roads, city walls, etc., and the changing terrain of the city as land was reclaimed for building and habitation. The developed methods and software can be of immediate use to the Newcastle City Planning Department's Archaeology Unit and other historians, but are also potentially useful for other similar organizations, such as architects, the police and the armed forces.

From this basis the project will allow archaeologists and other users to explore an accurate virtual historic environment with full freedom of movement. This allows their understanding of the model and the data to be enhanced by fully immersing themselves in the data set. An example of this is the ability to enter houses and look out through windows. This will give a greater understanding of the location of monuments and the lifestyle of past inhabitants. Further organization and spatial cognition is added to the data set in line with understanding of the world as it is viewed now. The data set is divided into separate sources, both digital and paper, and is inherently two-dimensional in its nature. In the process of creating the environment, the major part of the paper data will be converted into a digital format. This will enable it to be accessible from inside the three-dimensional model, to allow its display in a multimedia form using modern visualizer technology.

The historic nature of the model poses the recurring problem that there is only a single image – often either a photograph or a sketch – of many of the buildings in the study. Because of this, and the fact that we have no knowledge of the camera, or eye, that was used in its creation, it is difficult to use existing photogrammetric techniques to build the models. The solution to this has been approached by both the artificial intelligence community (Marill, 1995) and photogrammetrists (Newsam, 1996). The software being developed for this project builds on these ideas to allow for the semi-automatic recreation of three-dimensional scenes from a single two-dimensional image. This is something currently unavailable to the archaeology unit and will aid in the recreation of an historic 3D model of Newcastle upon Tyne.

7.2   DATA SET

A sample test area for the approach has been selected as the quayside area of Newcastle upon Tyne, around the mid-seventeenth century. This area offers interesting relief, with a steep slope rising from the banks of the River Tyne, together with a rich variety of approaches for the creation of models of archaeological monuments. The Archaeology Unit maintains a substantial archive of photographs and other images of buildings in and around this location. A number of buildings from the era still exist and will be used during a trial of various methods, suggested below, for model reconstruction. Statistical tests will then be carried out on the resulting models in order to ascertain the relative accuracy and benefits gained from the differing techniques.

7.3 MODELLING THE LANDSCAPE

Data defining the changing landscape of the city has been collected from various sources including the Ordnance Survey, the British Geological Survey and the work of other archaeologists. The landscape data has been processed by the Quicksurf surface-modelling software inside AutoCAD, a desktop, computer-aided design package. Historic mapping, by Thomas Oliver, circa 1830, has also been digitized into the system to allow for its usage as a drape image for the landscape and for the accurate positioning of the buildings. The data set available covers an area of approximately 2 000 m by 1 500 m. A triangulated grid of 5 m has been interpolated across this area using Quicksurf. In order to aid in later processing and display, this grid was sectioned into 100 m^2 tiles. The whole landscape was then exported out of AutoCAD in the Drawing Interchange File (DXF) format and into the virtual reality software, Superscape. This process involved a number of difficulties owing to the assumptions made by the Superscape software. These are that no tiles overlap, and that the triangular faces were created in a clockwise manner. Failure to observe these rules will produe detrimental affects on the display of the landscape. Substantial editing will then inevitably be required.

7.4 MODELLING THE BUILDINGS

Newcastle City's digital, urban, archaeological record holds the geographical location and the description of all the archaeological finds within the city. Furthermore, sketches, photographs, engravings and paintings of the historical structures have been converted into a digital form by scanning, to allow for their use in the creation of the three-dimensional models. Finally, it is hoped to model in detail the interior of the buildings from archaeological knowledge of the period and lists of the contents of buildings obtained from their previous inhabitants' wills.

The buildings are being modelled by a variety of methods. Those still standing can be surveyed accurately by the use of three-dimensional, digital survey techniques using conventional total station and reflectorless electronic distance measurement (EDM). The project will use the standard method of simply capturing the corners of building faces and will also try the innovative approach of capturing points on the faces and recreating the structure from the resulting face intersections (Stilwell *et al.*, 1997). As an alternative approach, photographs of these existing buildings can be taken and photogrammetric techniques used. These methods depend on the type, quality and quantity of images obtained for a single building. Methods being addressed by the project are stereo correspondence between pairs of images, and computer-aided models from multiple or single oblique images.

7.5 PHOTOGRAMMETRIC MODELLING

Interest in the extraction of 3D shape from 2D imagery has grown in past years from various fields of science. It has moved from a mainly photogrammetric approach to one involving the disciplines of artificial intelligence, machine vision, virtual reality and multimedia graphics. This has arisen from the use of robotics in the work place and a move towards 3D data structures, as opposed to previous plan views of environments.

Historically, the science of photogrammetry deals mainly with photographs that have a known position and orientation at the time of exposure, and are usually taken by a metric camera. Today we see a move towards the use of non-metric cameras – for example, standard off-the-shelf 35 mm models and little, if any, knowledge of the scene at time of exposure. This lack of knowledge appears most in the use of historical imagery taken purely for illustrative purposes. Another change is from using stereo pairs or multiple images of the same object, to the use of a single image or video images. This is due to the increase in the speed of data capture using these methods, and the higher availability of this type of imagery.

New methods, such as the one described by Streilein (1994), have advanced the ability to construct a three-dimensional model at least semi-automatically, if not fully automatically. The amount of user input can depend on various factors – for example, the information supplied about the image, the required accuracy of the final model and the objects within the model. One of the main theories for use with architectural models is a combined image- and geometry-based approach to the reconstruction.

For the 3D shape of an object to be restored from a single image, it is necessary to recreate the camera's position and orientation as they were during the exposure of the image. This conversion from 2D to 3D occurs automatically in the human brain nearly every second of our lives. Knowledge applied to the image received by the brain from the eyes allows for the interpretation of the scene. A large number of objects are possible in the 3D world to match a single 2D image. Thanks to facts learnt about objects and image formation when young, a best fit model can be derived. Possibilities still exist to confuse the human brain, as is seen in the case of optical illusions.

By identifying some of the assumptions made by the brain about objects in three-dimensional space, it is possible to begin to recreate models from images computationally. These assumptions include geometrical constraints on the objects in the image, which can be mixtures of angles and distances between set points and planes on the image. This method allows some of the unknowns to be resolved: for example, a camera will have six degrees of freedom and an unknown principal distance.

Debevec *et al.* (1996) and Newsam (1996) both present a combined geometric and photogrammetric approach to image reconstruction when applying techniques to reconstruct architecture in imagery. This is done by the user illustrating on the image various geometric constraints about points, lines, surfaces and volumes. Then, at any time during this process the user can create the model depending on these parameters and in this way forming a best fit model to the parameters supplied by minimizing errors in the geometry.

7.6   SINGLE IMAGE MODELLING

Prototype software being developed by the department uses the idea of image-based and geometry-based modelling techniques. The starting-point for the software is the functional model of the central perspective projection (see Figure 7.1 below). This applies to the unique case of a perfect pinhole camera, with its lens located at the perspective centre, O. For all the imagery used in the project this perfect projection will not exist, and the lens distortion must be taken into account for an accurate model to be constructed from the image.

Cooper and Robson (1996) show that the point A, on the three dimensional model, is projected on to the image plane by the straight line AOa. The relationship between the two points, A and a, can be represented by:

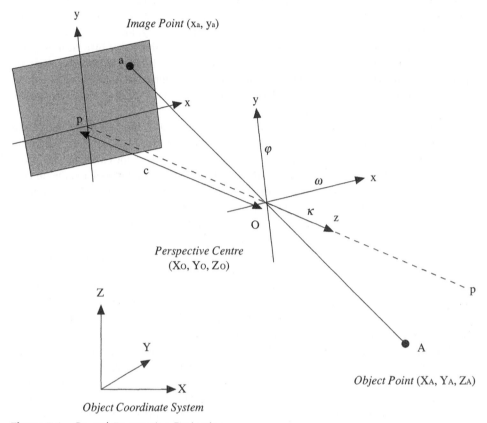

Figure 7.1   Central Perspective Projection.

$$X_A = X_O - \mu R^t x_a \quad \text{and} \quad x_a = \mu^{-1}R(X_O - X_A) \tag{7.1}$$

In matrix notation these equations can be represented as:

$$\begin{bmatrix} X_A \\ Y_A \\ Z_A \end{bmatrix} = \begin{bmatrix} X_O \\ Y_O \\ Z_O \end{bmatrix} - \mu R^t \begin{bmatrix} x_a \\ y_a \\ -c \end{bmatrix} \quad \text{and} \quad \begin{bmatrix} x_a \\ y_a \\ -c \end{bmatrix} = \mu^{-1}R \begin{bmatrix} X_O - X_A \\ Y_O - Y_A \\ Z_O - Z_A \end{bmatrix} \tag{7.2}$$

where R is the matrix of rotations for the camera orientation:

$$R_{\omega\varphi\kappa} = \begin{bmatrix} \cos\varphi\cos\kappa & \sin\omega\sin\varphi\cos\kappa + \cos\omega\sin\kappa & -\cos\omega\sin\varphi\cos\kappa + \sin\omega\sin\kappa \\ -\cos\varphi\sin\kappa & -\sin\omega\sin\varphi\sin\kappa + \cos\omega\cos\kappa & \cos\omega\sin\varphi\sin\kappa + \sin\omega\cos\kappa \\ \sin\varphi & -\sin\omega\cos\varphi & \cos\omega\cos\varphi \end{bmatrix} \tag{7.3}$$

The values ω, φ and κ are rotations of the camera viewpoint about the x, y and z axis respectively. The rotation matrix R is orthogonal, so $R^{-1} = R^t$. The value μ is a scalar quantity greater than zero, giving the scale of the relationship between the image and the model. In the case of a photograph where the projection plane is on the same side of the perspective centre as A, it is necessary for c to become positive. It is important to note that the image co-ordinates of a point, x_a and y_a, rely on the location of p, the principal point of the image, which may not always be the centre of the image.

It is, therefore, first necessary to solve the seven unknowns for the camera and then use this information to calculate the position of labelled features in the image. These seven unknowns consist of: the three translations for the position of the camera in the object co-ordinate system, X_O, Y_O, and Z_O; the three rotations for the position of the camera, ω, φ and κ; and the principal distance of the camera, c (refer to Figure 7.1).

The distribution of vanishing points is based on the camera position relative to the object; it is, therefore, possible to recreate the position of the camera from the known vanishing points. The vanishing points are inputted manually by use of selecting parallel features on the image – for example, the top and bottom of windows and buildings – and then calculating where the construction lines through these intersect. Automatic algorithms do exist for the extraction of vanishing points from an image – for example, Strafoni et al. (1993) and Parodi and Torre (1994). In the current software no attempt has been made to implement automated methods of vanishing point creation, as user intervention is more accurate at the current time.

The number of vanishing points on the image relates to the number of axes of the object co-ordinate system which the image plane intersects. A single image can therefore yield 1, 2 or 3 vanishing points (referred to here as 1-VP, 2-VP and 3-VP). For a building there often exists 3-VP, although a horizontal view down a street may only yield 1-VP. If an image has 2-VP or 3-VP, it is possible to reconstruct the camera position. Where there is only 1-VP, it will be necessary to consider alternative methods to reconstruct the model. Figure 7.2 shows an example of the imagery being used in the project, illustrating both the quality and orientation of a typical view. The vertical walls of the buildings appear to be parallel, thus implying only 2-VP, although this causes the horizontal line not to fall through the principal point p, defined as the point where the z axis intersects the image plane (normally the centre of the image). This may be due to the fact that the image has been cropped, or that the third vanishing point falls close to infinity, suggesting only a slight tilt in the image plane. Further testing is planned to see what differences occur in treating an image which has a third vanishing point close to infinity as if it has only 2-VP.

Zhao and Sun (1994) state that an image with 2-VP has its x axis along the line passing through the two vanishing points V_x and V_y. Lines parallel to the X axis will all intersect at V_x and similarly lines parallel to the Y axis will all intersect at V_y. In the case of 2-VP all lines parallel to the Z axis are parallel on the image.

The perspective centre O, will then lie on a semi-circle which has a diameter of V_xV_y at a point perpendicular to the principal point p, of the image, as shown in Figure 7.3.

The geometry of a circle states that two points on opposite sides of a circle always form a right angle when joined by a third point, giving OV_x and OV_y to be perpendicular to each other. The view line from O to the point V_x must be parallel to the X axis because its perspective projection will also be through the point V_x. By the same argument the Y axis can be shown to be parallel to the line OV_y (the Z axis is defined as in a right-hand co-ordinate system, therefore parallel to the y axis in the 2-VP case). The principal distance c, the distance from the perspective centre O to the principal point p on the image plane, can then be calculated by;

$$c = |Op| = \sqrt{(|x_{vx} - x_p| * |x_p - x_{vy}|)} \qquad (7.4)$$

For Figure 7.2 the principal distance of the image has been calculated to be 116 mm, assuming 2-VP, for an image size of 130 mm by 95 mm. The accuracy of this measurement

Lower Pilgrim Street, Newcastle (Photograph by Mr Edwin Dodds)
Proc. Soc. Ant. Newc., 4 Ser. III.

Figure 7.2 Example image.

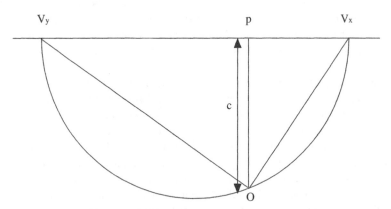

Figure 7.3 Geometry of two vanishing points.

relies on the accuracy of the position of the two vanishing points and the principle point of the image.

By choosing the origin of the object co-ordinate system to be a point in the image plane – for example, the bottom corner of a building – it is possible to get relative object co-ordinates and rotations for the camera, thus allowing for the reconstruction of the model by the equations detailed earlier. The co-ordinates are later transformed into absolute co-ordinates by fixing the model to its position on the historic map.

7.7 FUTURE WORK

Further intended work includes using the least squares method to report back the relative accuracy weighting of any given point, line or face. Where the image is in the form of a sketch, information is given a lower accuracy weighting and then altered and refined during model development. The model will constantly update as new primitive objects and constraints are added to enable the most accurate final result.

Other planned developments include: experimentation with the selection of single points from the image, such as those identified for calculation of vanishing points. The consequence of varying these on the final model will be investigated; direct comparison between the model and actual relative co-ordinates, for linear and areal features; least squares adjustment of derived angles and distances; the adoption of a technique to determine positional accuracy of linear features, proposed by Goodchild and Hunter (1997); an adaptation of the use of homogeneous co-ordinates and Gaussian mapping to represent the orientation of lines on a unit sphere, in order to compute vanishing points, as described by Barnard (1983) and Leung et al. (1997).

The final model will be created in a basic form, giving it the ability to be utilized by a variety of existing and future visualization packages and techniques. It is important that the methodology can be applied and visualized as easily on a high-powered workstation as it can on an entry level personal computer. The use of emerging standards for the dissemination of information over the Internet, such as the virtual reality modelling language (VRML), Java and ActiveX, will provide the opportunity for all those interested to interact with such models as the one being developed in this project. This methodology utilizes existing, affordable, three-dimensional and virtual reality software in the creation of the visualization.

The collaborating archaeology unit are seeking to extend the implementation of these techniques for the whole of the Newcastle City Urban Record area. The major objective is to create a 'virtual medieval town' in a permanent installation in Newcastle upon Tyne. It is envisaged that this will be used as a major tourist attraction, with which visitors can instructively explore an ancient city.

7.8 CONCLUSION

This particular project confronts the problem of creating accurate three-dimensional models by interpreting single perspective images. The methods used and described here are an effort to integrate procedures and algorithms developed in research disciplines other than geomatics, such as artificial intelligence and machine vision, to the requirements of accurate, large-scale data capture. It is obvious from the literature review undertaken so far, that there are many other computational techniques being developed for the interpretation and identification of geometric properies in images that are of relevance to accurate three-dimensional modelling. Geomaticians have for a long time been able to make basic measurements in three-dimensional space. The tools are now available to realize the full potential of these measurements. The creation of accurate, full, three-dimensional models of the real world, which can be interactively explored, investigated and analyzed from any position in space is now all possible. This project is part of a continuing programme of investigations being undertaken by the Department of Geomatics into the use of state-of-the-art technological tools for the creation, manipulation and analysis of accurate, large-scale survey data in true three dimensions.

References

BARNARD, S.T. (1983) Interpreting perspective images, *Artificial Intelligence*, **21**, 435–462.

COOPER, M.A.R. and ROBSON, S. (1996) Theory of close range photogrammetry, *Close Range Photogrammetry and Machine Vision*, Atkinson, B. (Ed.), 9–51, Latheronwheel, Scotland: Whittles.

DEBEVEC, P.E., TAYLOR, C.J. and MALIK, J. (1996) Modelling and Rendering Architecture from Photographs: A hybrid geometry- and image-based approach, *Conference proceedings of SIGGRAPH '96*.

GOODCHILD, M.F. and HUNTER, G.J. (1997) A Simple Positional Accuracy Measure for Linear Features, *International Journal of Geographical Information Science*, Vol. 11, **3**, 299–306.

KRAAK, M.J. (1989) Computer-assisted cartographic 3D imaging techniques, in J. Raper (Ed.), *Three Dimensional Applications in Geographical Information Systems*, 99–114, London: Taylor & Francis.

LEUNG, J.C.H. and MCLEAN, G.F. (1997) Vanishing Point Matching, http://www.me.uvic.ca/~jleung/icip96.html.

MARILL, T. (1995) The Three-Dimensional Interpretation of a Class of Simple Line-Drawings, *A.I. Memo 1555*, Cambridge, Mass: Artificial Intelligence Laboratory, MIT.

NEWSAM, G. (1996) SolidFit – Shape and Measurements from Perspective Distortion, http://www.cssip.edu.au/~vision/solidFit.html.

PARODI, T. and TORRE, V. (1994) On the Complexity of Labeling Perspective Projections of Polyhedral Scenes, *Artificial Intelligence*, Vol. 70, 239–276.

PETCH, J. (1994) Epistimological aspects of visualization, in M. Hearnshaw and J. Unwin (Eds), *Visualisation in Geographical Information Systems*, 212–219, Chichester: John Wiley.

STILWELL, P., TAYLOR, G. and PARKER, D. (1997) Precise Three-Dimensional Measurement and Modelling of Structures Using Faces, *Surveying World*, Vol. 5, **3**, 31–33.

STRAFONI, M., COELHO, C., CAMPANI, M. and TORRE, V. (1992) The Recovery and Understanding of a Line Drawing from Indoor Scenes, *IEEE Transactions on Pattern Analysis and Machine Intelligence*, Vol. 14, **2**, 298–303.

STREILEIN, A. (1994) Towards Automation in Architectural Photogrammetry: CAD-Based 3D-Feature Extraction, *ISPRS J. Photogrammetry and Remote Sensing*, Vol. 49, **5**, 4–15.

TEMPLFI, K. (1997) Photogrammetry for urban 3-D GIS, *GIM*, Jan. 1997, 60–63.

ZHAO, X. and SUN, J. Reconstruction of a Symmetrical Object from its Perspective Image, *Computers and Graphics*, Vol. 18, **4**, 463–467.

Artificial Intelligence, spatial agents and fuzzy systems

PART THREE

Artificial Intelligence, spatial agents and fuzzy systems

CHAPTER EIGHT

How to deal with water supply pollution with GIS and multi-agent systems (MAS)

CHRISTIANE WEBER

8.1 INTRODUCTION

Among fundamental needs, water is still essential. Human beings cannot live without it and this provides water with a prime value. The global need worldwide is evaluated at 5 200 km^3 per year (which means 5% of all the precipitations over land masses), divided into 65% for irrigation, 26% for industrial and 9% for domestic use. To give some idea of the volume of domestic use, the consumption of water for the planet is estimated at 263 km^3 per year, let's say 140 litres per day per inhabitant. A Parisian needs around 250 litres of water per day. Such water is potable at the point of supply, but is returned polluted in the sewage system. Water is considered as a convenient vector to eliminate human waste. Consequently water use carries many implications: political, economic, cultural and technical. Nowadays, a general trend in developed countries' attempts to extend water use from priority needs towards new uses like leisure, swimming, local initiatives . . . and in the same way environmental concerns have emphasized the necessity of controlling the balance, regulation and cycle principles related to the water system, in order to preserve water resources. Such concerns have led to policies on water use, with the stress particularly on (1) water quality according to supply, use or production, (2) overground and underground water protection associated with used water, (3) technical needs for water treatments.

Looking at the urbanization trends all over the world, it is easy to see that water resource is a major concern for urban authorities. They have the responsibility to the public of guaranteeing water supply in quantity and in quality. Difficulties appear when considering the water cycle because of the complexity of the different stages of the cycle, the natural characteristics of the site and the influence of the different elements (Figure 8.1).

In different countries, the precise areas of responsibility between state and local authorities have been defined, creating water management organizations (for example the SAGE in France – Schéma d'Aménagement de Gestion des Eaux) and specifying the powers of

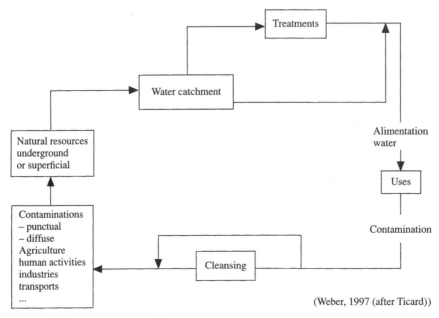

Figure 8.1 Schema of water cycle.

intervention of urban authorities in accordance with the World Health Organisation (WHO) safety standards, the legal aspects of the protection of water sources, and so on.

In this respect, the present project deals with water supply security in cities with a view to analyzing their capabilities for managing a risk situation – for instance, the spread of a pollution substance through the water resource.

The remainder of this paper is structured as follows: First the presentation of the systemic approach which is used to reduce the complexity of the situation. Secondly, the methodological approach is explained with the needs and the developments required. Thirdly the experiment is presented, opening up some discussion.

8.2 THE SYSTEMIC APPROACH

The complexity of the reality studied may be summarized by a systemic approach. In fact several systems could be defined (Table 8.1).

Table 8.1 Systems presentation.

Water resource system	Socio-economic system	Legal system	Technical system
Ground table	Users: Population and companies	Protection authorities and infrastructures	Distribution net (collecting and distribution)
Rivers and channels Gravel-pits . . .	Management organizations		Sewage net (recovering and treatments)

If we consider the water resource, the *physical system of general water circulation* may be described by its different components: the water-table, the rivers, the channels, the gravel-pits, ... and the flux between them. The water-table, in this study, corresponds to the essential source of water for Strasbourg. In the case of attack by a pollution substance, depending on its severity, the contamination situation could give rise to concern.

> **Water-table:**
> length: 20 km, width: 50 to 100 m;
> speed: several metres per day; catchment per day 130 000 m^3;
> quality: drinkable without treatment;
> cost 10,13 F/m^3 (national: 12,55 F/m^3).

The second system involved is the *economic and social* system. The city as a whole could be represented by the different uses of water in a social organization. Uses relate to the water characteristics – as driving force, as transportation, for heating, as a drinkable water supply or as a support for leisure activities. Different categories of activity are involved, increasing the complexity of the system. In our case study, we focus only on the *domestic* and *industrial* use, including the distribution of drinking water and its networks management and users.

The third system, associated with the previous one, is the *legal responsibility for the protection of the resource*. This was defined in the 1970s: local and national government share the setting of the protection perimeter. The national policy guidelines have been reinforced by European directives regarding the protection perimeter necessity. This perimeter is determined according to different distances from the abstraction point in order to avoid prejudicial locations (industrial activities or residential location). At the end of 1997, all the pits in Europe must be associated with a protection perimeter (divided into three zones for the majority of European countries, distance being calculated in metres and days) which may have serious consequences for older ones, owing to the spread of urbanization or agriculture. One point to be noted: this protection takes care only of the surface water, no directives concern specifically the ground water. Some concessions may be obtained – for example, a pipeline can cross the perimeter. The protection infrastucture is also determined by piezometers sets with manual or automatic records.

The fourth system, which is also associated to the social one, is the *technical* one which has three functions: water distribution; sewage networks; and post treatment infrastructure.

In order to reduce the complexity of the situation studied, we may define a sub-set system, the *water supply use system* (Figure 8.2) to describe the relationships between the source (water-table) and the different groups involved in water distribution: the user's and the management and protection authorities. Thus the function outlined here is the supply of water in quantity and quality according to the user's requirements.

Quantity and quality criteria are important factors to be taken into account. Quantity involves the control of water demand and the anticipation of possible problems of distribution. Quality is perhaps more difficult to guarantee because of the quality standards which are in force. If we consider only the drinkable water supply, rigorous health standards are involved and numerous tests are carried out. Piezometers measurements are continuously recorded to be analyzed by local authorities, on the one hand, and by health ministries' local representatives, on the other. Of course, there is communication between these two bodies.

The subject of the study is based on the analysis of a risky situation disturbing water distribution. The *main hypothesis* of this study[1] is based on the inadequacy of the intervention rules (due to the linearity of actions) related to the spatial and temporal characteristics

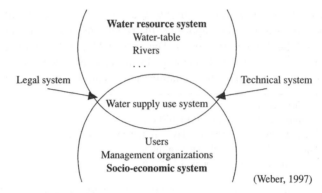

Figure 8.2   Water supply use system.

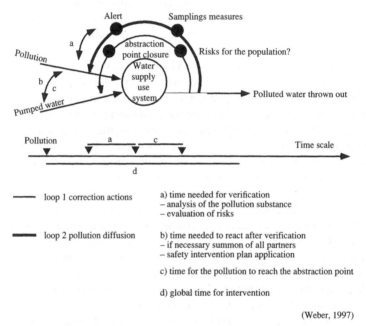

Figure 8.3   The water-supply-use system processes.

of the phenomenon to control, for instance the pollution front (Figure 8.3). The accidental spread of a pollution substance could result in a series of effects able to disturb, possibly for a long period, the water-distribution process. The delay between the spread of the pollution and its detection is crucial for the on-going mechanisms. Depending on the visibility of the pollution (a lorry spillage, for example) or the lack of it (industrial infiltration, for instance), the reaction loop will be quick or slow and more, or less, appropriate. If some slowness does affect the reaction process, distribution may be disrupted. This may have consequences for users and, possibly in an irreversible way, on the water-supply technical network. A disruption of the water supply may even endanger life.

It is necessary to define what could be a risky situation for water distribution. We can identify two kinds of pollution: a *diffuse contamination* usually linked to modern farming methods (nitrates, and so on) which may reach the water-table at several locations, and *instant contamination* due to a precise event (accident, breaking of a circuit, . . .). The first is difficult to handle in such a study because it would need to operate at the general water-table scale, which means the French part of all the Alsatian plain. The second may be simulated to cope with a feasible event (such as a traffic accident).

In reality, in order to get to grips with the problem, we need to study two cases: a *normal* water cycle – to establish a diagnosis of the water-supply use system, its component states and their relationships; and an *abnormal* one describing the disturbed system and what is needed to restore the normal distribution situation.

8.3 METHODOLOGICAL APPROACH

In order to study the two cases previously mentioned, several steps have to be specified:

- The *diagnosis of the water-supply-use system*, through the components and the relationships description, which relies on the information we can collect and structure. Different sources have been identified: the city services, the Bureau de Recherches Géologiques et Minières (BRGM), the Mécanique des Fluides Laboratory and the local Health Ministry service (Direction Départementale de l'Action Sanitaire et Sociale). The simulation site is located in the Robertsau well area, which has been chosen because of the location of industrial activity nearby (i.e. petrol stations, a busy harbour, a pipeline, and a number of factories – paper, cleaning, painting, etc.), the independence of the water distribution net from the general network, the priority population to be served (hospital, nursery, schools, old people's homes etc.) and the interest in this city, a study is on the way to open another well in this area.

- The *elaboration of the accident scenario*: according to the location of the study the *accident simulation* could have been of three types, if we consider only the hydrocarbon risk – a leak in the pipeline, an overflow of petrol in a tank or a lorry accident on the road near a petrol station. The last example has been chosen, after discussions with the authorities, as it seems the most realistic. The simulation follows the rules of the traffic scheme for dangerous transportation.

    A similar problem appeared in 1991, when tetrachloroethylene was discovered at a boring well outside Strasbourg. The pollution was traced to an industrial site nearby. A depollution site rapidly helped to restore safe water quality. This kind of pollution is associated with the chemical industry and its products, and used by industrial, craftsman or domestic activities. Heavier than water, the products are dispersed into soil, rivers and water-tables.

- The study of the pollution management scenario requires the examination of the different steps to be taken from the initial discovery to the conclusion of the crisis. This may be effected by the analysis of the roles of the organizations involved and their relationships, and the series of activities which need to be triggered. This step is important because it makes it possible to identify the *geographic entities* which are needed and the links they possess regarding the event they are describing, the *roles of the organizations and the relations* maintained during the duration of the pollution crisis.

■ The adjustment of the pollution management to deal with the propagation of the pollution substance is studied on a *timescale* corresponding to the duration of the carrying of the pollution by the water-table flow.

8.3.1 Methodological needs

To carry out this study we need to model the complexity of the water-supply-use system through the information collected, and to keep the essential dynamics of the situation the simulation approach is essential.

GIS capabilities are required to gather information spread throughout different bodies. GIS allows us to locate and identify the systems studied (physical resource, socio-economic, legal or technical), and the geographic entities which are involved in the two cases previously presented.

The complexity of the interactions between the social and the physical systems requires the use of the new concepts developed in the field of multi-agent systems. Classical artificial intelligence (AI) deals with a single agent of high complexity manipu-lating explicit knowledge of the domains. In contrast multi-agent systems not only focus on agent architectures but also stress the importance of interactions between agents. Depending on the kind of agents, multi-agent systems can be applied to various domains for simulation or problem-solving. Simple reactive agents allow us to focus on the con-cept of the emergence of global phenomena from interactions between agents and with the environment (Ferber and Muller, 1997). In our case it should be possible to handle the system's dynamics' characteristics with multi-agent system-simulation. More sophisti-cated cognitive agents allow us to model social organizations and their dynamics. More-over, the linking concept of interactions allows us to model the various points of view and link them together from the physical to the social level. This aspect is managed by the Artificial Intelligence Laboratory of Neuchatel (CH).

8.3.2 Methodological developments

To cope with these requirements we have to model the situation both in a normal and in a crisis situation, highlighting the *pollution propagation* modelling in the water-table and the *organizational modelling* required to meet it.

Through our partnership with the Mécanique des Fluides Laboratory we have the results of the pollution propagation showing the duration of the journey and the proposed accident locations. We decided to follow the 'worst case' scenario for pollution, because we assume that for such situations the crisis organization scheme will identity possible deficiencies. Thus the speed of hydrocarbon pollution is considered to be the maximum, equal to the water one. For the same reason only the pollution front will be kept, no mention of the possible dilution (no deposits or recapture mechanisms will be taken into account) are held. We consider that all the water affected by the pollution will be unsuit-able for domestic use.

To model the organizational aspects we base our development on real cases which have taken place in France (Lyon in 1988 and Tours in 1992) in order to work out action durations and to analyze the organizational relationships using official documents.

Table 8.2 List of the collected data.

Systems	Nature	Form	Scale	Source	Date
Physical	piezometry	maps	100 000	BRGM	1994
	quality	maps	100 000	BRGM	1994
	nitrate	maps	100 000	BRGM	1994
	sulphate	maps	100 000	BRGM	1994
	chlorure	maps	100 000	BRGM	1994
Technical	Wells	maps	10 000	CUS	1978
Water Alimentation	catchment field	maps	50 000	BRGM	1994
	protection perimeter	maps	5 000	CUS	1978
	water quality	tables		BRGM	1994
	debi?	table		CUS	1995
	net A.E.P.	maps	1 000	CUS	1980
		maps	4 000	CUS	1995
		maps	20 000	CUS	1994
		maps	4 000	CUS	1995
Socio-economical	Priority subscribers	table		CUS	1996
	risky activities	tables + maps	10 000	BRGM	1995
	– pipe-line	maps	80 000	CUS	1995
	– circuit TMD	maps	50 000	SPPPI	1994
	– sewage net	maps	25 000	CUS	1994

8.4 EXPERIMENTATION

The study relies on a body of information gathered in a GIS project describing the geographic site (the river infrastructure, the road network, housing . . .) and the water resource system, bearing in mind that the water-table runs completely under the city. We decide to use the model calculated by the Mécanique des Fluides Laboratory and to keep the temporal and spatial nature of the pollution transfer system. The socio-economic system has been described by the geocoding of the priority population, of the potential risk societies, the geocoding of geographic entities (pipe-line) or location (petrol station). The legal system is described by the protection perimeter. The protection system corresponds to the identification and the location of the piezometers and the perimeter around the abstraction point. The distribution net is introduced in the GIS with the abstraction point and the connecting network (Table 8.2). As can be seen, heterogeneous data have to be integrated in order to get the relevant information. Different spatial analyses specify the relationships between entities, particularly the population affected by a possible closing of the abstraction point, or the location of necessary piezometers according to the water flow and the existing or planned abstraction points.

Figure 8.4 shows the steps of the management water supply in each case (without and with disturbance)[2]. It is clear that some bodies will have more than one role, and of course more than one action to undertake. The analysis of the decisional process is based on some hypotheses which need to be explained: first we assume that nothing (or almost

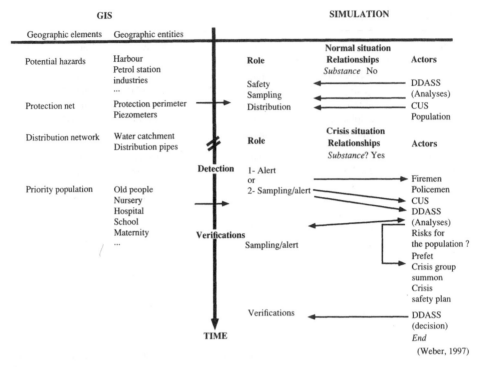

Figure 8.4 Water management.

nothing) in the process is unpredictable, except the accident. The reason for this is the reproductibility of the results which are assumed to be realistic. For the same reason, we consider that the actual behaviour of the organization is quite close to the theoretical behaviour, so we do not take into account personal initiatives.

To summarize the planned simulations: we have four entities for the hydrodynamic simulation: the water-table, the abstraction point (for the distribution), the measurement samplings (to verify the quality) and the disposal of the pollution substance. All these entities have spatio-temporal behaviours.

For the organizational decision simulation we need to identify what could be the agents of the system. We prefer to consider the agents under *objectives definition* like water alimentation, quality control, prevention process and crisis-management. All these objectives have, of course, their own spatial competences, – local, county, national or European. What has to be defined are the organizational schemes (role plus relations) in which the different organizations are involved in order to define the closure of the simulated system and its coherence. *Three* aspects might be emphasized for tackling this problem:

- Distribution objective.
  Role (management, service, responsibility).
  Relation: network management, monetary flux, obligation towards the subscribers.
- Prevention and control objective.
  Role (control, detection, correction).
  Relation: quality control, alert, verification, starter, flow of information.
- Crisis management and termination.
  Role (identification, pollution source location, pollution reduction).
  Relation: identification, integration of information, starter of actions.

The simulation steps in the early stages and the computational part is just being developed. A prototype will be designed for this summer. Some tools are planned to facilitate use. Two types of tools may be distinguished, one corresponding to the model conception and the other to the exploitation of the results – for instance, spatial dynamic queries on the pollution transfer or the ongoing decisional process on the pollution timescale. These kinds of tools might provide a set of solutions which may be tested on different areas of Strasbourg or elsewhere.

8.5 CONCLUSION

This experiment does highlight the complementarity between GIS and dynamic modelling. The necessary, reality-reduction approach allows us to specify entities for such links. Of course a pollution event might be more complex than the study case proposed. But we assume that reproductibility might be feasible if we bear in mind the need to describe the systems, the entities and their relations. One of the difficulties is to connect the intervention process to reality because little information exists in this domain. For instance, 'time' is a data which may vary a lot from one situation to another and between the different roles involved, even if the management role is relatively easy to obtain. The emergent concept is particularly attractive in such a context, because it attempts to reproduce human actions for which roles, relations and organizational schemes do not accord exactly, most of the time, with strict directives.

Notes

1   This study corresponds partly to C. Ferron thesis.
2   Other organizations may be solicited for an accidental pollution, for example the services of the industry ministry when 'unknown pollution' is detected.

References

FERBER, J. and MULLER, J.P. *Formalizing emergent collective behaviours: preliminary report*, DAIMAS '97, St Petersburg, June 1997.

FERRON, C. (1994) *Mobilisation des connaissances et définition de stratégie d'action: l'information géographique au service de la compréhension et de la gestion des ressources et des milieux naturels*. Programme environnement du CNRS.

FERRON, C. (1996) *Sociétés et hydrosystèmes: difficultés d'application des procédures de protection des puits de captages; application géographique et modélisation*. Journées GIP Hydrosystèmes.

LABBANI, O., MÜLLER, J.P. and BOURJAULT, A. (1996) *Describing collective behaviour*. ICMAS '96.

MINISTÈRE DE L'INTÉRIEUR ET DE L'AMÉNAGEMENT DU TERRITOIRE (1993) Direction de la sécurité de la sécurité civile. *Pollution accidentielle des eaux intérieures*.

MÜLLER, J.P. and PECCHIARI, P. (1996) *Un modèle de système d'agents autonomes situés: application à la déduction automatique*. JFIADSMA '96.

CHAPTER NINE

Some empirical experiments in building fuzzy models of spatial data

LINDA SEE AND STAN OPENSHAW

9.1 INTRODUCTION

The electronic and spatial data revolution of the last two decades has resulted in the development of a variety of data-management tools for handling the vast amounts of temporal and georeferenced data that are being created. The advent of GIS marks a major step forward in integrating multiple layers of spatial information from multiple sources. However, this technology is very deficient when it comes to coping with the subsequent need for spatial analysis and modelling, and there is a noticeable absence of suitable and relevant technologies. One reason is the slow recognition that human systems modelling is of an order of magnitude more complex than classical statistical and conventional mathematical modelling methods can readily handle. The lack of adequate, prior, theoretical knowledge has resulted in the growing failure of quantitative geography to do much, or indeed anything, with most of the data being created by GIS other than to map it. Additionally, real doubts may be expressed about whether the prevailing Aristotelian-logic-based paradigm of science provides an adequate framework for spatial science (see, for example, Burrough and Frank, 1995). The existing crisp, logic-based science paradigm may well be too limited to represent real-world spatial phenomena adequately, and alternative texts, languages and representations may need to be considered. One route is to dabble with alternative non-scientific philosophies and ways of thinking (e.g. post-Modernism), but there are likely to be major practical problems with the incorporation of these ideas into GIS. Another approach, and one which would fit more easily in a GIS framework, is to investigate whether there are more scientific alternatives that can cope with the complexity of geographical phenomenon without appealing to the paranormal for inspiration.

It is perhaps fortunate that at more or less the same time as conventional quantitative geographic technologies started to fail when faced with masses of spatial data, other major tool-kits based on artificial intelligence technologies have become available for use within geography. Some of these new computational methods are well suited to large and noisy datasets, and they offer distinct advantages over classical mathematical and statistical methods by being inherently non-parametric and non-linear. One type of AI-based technology that has been inspired by the way we think and which can be used effectively in spatial modelling, is called 'fuzzy logic'. Rather than think in crisp well-defined concepts, we are able to reason with vague, incomplete and often conflicting information. If we are able to process vast amounts of information in this way, it should be possible to re-equip our models and computers with a similar capability. Although fuzzy logic and fuzzy set theory were developed over 30 years ago (Zadeh, 1965), it is only since the late 1980s that the subject of fuzzy systems modelling has developed into a practical and general-purpose modelling technology, which fits easily into the framework of the intelligent GIS and spatial decision support systems advocated by Birkin *et al.* (1996).

Research on fuzzy models of spatial phenomenon is still at an infant stage, although initial work on fuzzy spatial interaction has been undertaken (see Openshaw, 1996 and 1997; See and Openshaw, 1997). This paper describes, through empirical examples, how fuzzy logic can be used to build models of two types of spatial relationships: spatial interaction and unemployment. It also illustrates the potential of fuzzy logic modelling as a new tool for GIS and spatial-decision support systems. Existing statistical and mathematical models together with models developed via a genetic programming approach are used as benchmarks for evaluating the performance of the fuzzy models.

9.2 FUZZY LOGIC MODELLING

Fuzzy logic modelling is based on the mathematics of fuzzy set theory, the generalized version of classical set theory in which the notion of membership in a set is modified to include partial membership, i.e. values ranging between 0 and 1 (Zadeh, 1965). These fuzzy sets, or 'membership functions' as they are known in a modelling context, are analogous to the numerical variables in a statistical or mathematical model, and are referred to as 'fuzzy' or 'linguistic' variables. Linguistic variables are used in the natural language rules or IF–THEN statements that constitute the rulebase of a fuzzy model; these rules essentially map fuzzy input membership functions to fuzzy outputs. An inference engine, which uses an extension of multivalued logic or fuzzy logic, drives the model by execution of the rules in response to system inputs to produce the system outputs. The model type used in this paper is referred to as the Mamdani-type linguistic model (Jang *et al.*, 1997) and more specifically the standard additive model (SAM) developed by Kosko (1997). Kosko (1997) has also shown that all SAMs are able to approximate functions to a desired level of error by covering the solution with a series of rule patches. Where the patches overlap, the model adds or averages them; where the patches are large and gaps exists between patches, the model interpolates between them. The basic fuzzy SAM can, therefore, theoretically replicate the behaviour of an entropy-maximizing, origin-constrained, spatial interaction model or a model of unemployment. As fuzzy systems are not based on mathematical or statistical models of the system under approximation, they are, therefore, model-free estimators. As such, there is no need to know in advance the precise nature of the statistical or mathematical associations, and the researcher can concentrate on formulating the concepts.

Although expert knowledge can sometimes provide all the rules needed to formulate a fuzzy model, manual fine-tuning of the membership functions and rules weights can be a time-consuming process. Alternatively, knowledge can be mined directly from the data through various inductive methods, some of which allow the incorporation of existing knowledge directly into the initial fuzzy model formulation. In this paper, two different fuzzy modelling approaches are investigated: (1) a hybrid approach, in which existing knowledge is combined with inductive methods, and (2) an entirely data-driven approach in which the fuzzy model is specified and optimized by other AI techniques. By examining the rules and corresponding membership functions that result from the fine-tuning process, new insights into system behaviour can be realized. Moreover, the knowledge used in the initial fuzzy system formulation can be combined from more than one body of theory, allowing the possibility of integrating many different ideas about the system behaviour into a unified model, and opening up the prospect of entirely new types of geographical model.

There are a number of different inductive methodologies that can be integrated into the fuzzy model formulation process, including genetic algorithms (Cordón and Herrara, 1995), decision tree and production rule-based methods (Quinlan, 1993), and knowledge-mining techniques (Wang and Mendel, 1992). Genetic algorithms (GAs) are non-linear optimizers which operate under the principles of natural selection and survival of the fittest. The entire fuzzy model can be coded into a binary string, which is analogous to an individual. A random population of individuals is generated, and the fittest members are selected for breeding. Over many generations, a whole new population of possible solutions, which possess a higher proportion of the characteristics found in the fitter members of the previous generation, will be produced. It is also possible to optimize only a portion of the fuzzy model, such as the membership functions and the rule weights, and leave expert knowledge and other better-suited rule discovery methods to finding the optimal set of rules. Decision trees are used frequently in the areas of multivariate analysis, expert systems and machine learning to partition sample data into a collection of rules. The C4.5 machine learning software package (Quinlan, 1993) produces decision trees that can be re-expressed as production rules, which are generally easier to understand. Jang *et al.* (1997) suggest using decision trees to partition the input space into crisp rules, and then to fuzzify the rules directly into initial membership functions. The Wang–Mendel knowledge-mining approach (1992) was developed as a quick and simple way to generate fuzzy rules from a series of input–output data. A fuzzy rule is created for each input–output data point and assigned a degree, based on membership and any available expert knowledge, which can be used to determine which data points are critical and which may reflect anomalies in the data. The conflicting rules are then eliminated by creating a combined fuzzy rulebase, choosing those rules with the maximum assigned degree.

9.3 SOME EMPIRICAL EXPERIMENTS IN SPATIAL INTERACTION

Spatial interaction is concerned with modelling the geographical variation in the movements of people, commodities, capital, etc., between differing geographical locations. Embedded in the framework of an intelligent GIS, the models can be utilized to examine the impact of changes to the system in a 'what if?' modelling capacity and to facilitate in the planning process. Empirical experiments with a journey to work dataset illustrate the potential application of fuzzy modelling to spatial interaction in terms of a flexible model framework, satisfactory performance and ease of interpretation.

9.3.1 A journey to work dataset

The 73 by 73 journey to work matrix for the county of Durham is a classic spatial interaction dataset that has often been used as a benchmark for testing different spatial interaction models in the past (Openshaw, 1976). The fuzzy-origin-constrained spatial interaction model can be calculated from

$$T_{ij} = A_i F_{ij} \qquad (9.1)$$

where T_{ij} is the predicted flow from origin i to destination j $(i,j = 1, \ldots ,73)$, F_{ij} is the fuzzy spatial interaction model and A_i is the accounting constraint.

9.3.2 Fuzzy model formulation

Five different methods were used to build fuzzy spatial interaction models for the Durham data set using a combination of the aforementioned fuzzy model building techniques. In Method 1, the C4.5 induction software was used to find an initial rulebase and a set of initial membership functions for the input and output variables. The C4.5 induction software found 48 rules which were then used to produce six membership functions for cost and five for both attractiveness and number of trips, reducing the rule count to 30. In Method 2, the Wang–Mendel rule discovery method was used to find an initial rulebase, adopting the membership function partitioning from Method 1. In Method 3, the rules created by Methods 1 and 2 were combined into a single set of rules based on knowledge of spatial interaction, eliminating or adding rules where deemed necessary. A GA was then used to fine-tune the rule weights and membership functions to produce an optimised fuzzy model given each of the rulebases in Methods 1 to 3. In Method 4, a GA was used to optimize the entire model including the rulebase but with a restriction on the membership functions so that overlapping was minimized to include only immediate neighbours. Finally in Method 5, a GA was used to optimize the entire model, but was given complete freedom with regard to the overlapping of the membership functions. For Methods 4 and 5, cost and attractiveness were partitioned into three membership functions, while the number of trips was partitioned into five.

9.3.3 Model goodness of fit

The performance of conventional spatial interaction models ranges from a residual standard deviation of 20.7 with a power function to a performance limit at around 15.3 for more complex deterrence functions such as Tanner or Weibull. The best result achieved by a genetic programming approach is 11.2 (Diplock, 1996), which achieves significant improvement over conventional models, although the resulting equations generated by this approach are rather complicated to interpret. The results for the fuzzy models using Methods 1–5 are listed below in Table 9.1.

After fine-tuning with a GA, all the fuzzy models generally performed well in relation to the conventional models, with slight improvements in performance gained with Methods 1 and 4. Method 3 produced surprisingly good results using only the rule-discovery methods and a bit of knowledge to correct the rulebase; unfortunately, further fine-tuning by the GA produced a disappointing final result despite a promising start. However, the fuzzy model results are still lagging behind the best performance achieved with genetic programming although, given the right combination of fuzzy-model building techniques and the optimal number of fuzzy sets, better performance should be, theoretically, possible. Fuzzy spatial interaction models may also be far more robust in scenario-testing because of their generalized structure.

Table 9.1 Residual standard deviations for the fuzzy spatial interaction models.

Method No.	Fuzzy Model Building Techniques	Residual Standard Deviation
1	C4.5 rule induction	31.0
	after fine tuning	14.2
2	Wang–Mendel	23.7
	after fine tuning	16.7
3	C4.5 + Wang–Mendel + knowledge	17.2
	after fine tuning	15.9
4	restricted total optimization	15.0
5	unrestricted total optimization	15.7

Table 9.2 Rulebase for the fuzzy model produced by Method 4.

Cost	Attractiveness		
	low	medium	high
small	some	many	massive
average	a few	some	more
large	(none)	(none)	a few

To illustrate the simplicity of the fuzzy models, Table 9.2 lists the final rulebase for the fuzzy model produced with Method 4, and Figure 9.1 depicts the final optimized fuzzy sets. Method 4 produced a good performing model with 3 membership functions for cost and attractiveness and only 7 rules, making it slightly easier to interpret than the models produced with Methods 1 to 3. Although the entire model was optimized with a GA, the restriction on overlapping between membership functions ensured a comprehensible set of final linguistic variables. The rulebase shows, in plain English, the expected pattern of larger numbers of trips with lower costs and higher attractiveness, decreasing as both cost increases and attractiveness decreases. The two bracketed values labelled *none* are crisp rules, in that the GA removed these rules from the rulebase; when these rules fire, they result in no interaction.

Now that fuzzy spatial interaction models have been shown to perform well on a benchmark dataset with easily interpretable rules and membership functions, the next step is to develop fuzzy models for larger and more complex systems of the kind that are dealt with in spatial-decision support systems such as retail and urban planning, location optimization, etc. The flexibility of fuzzy models also means that 'what if?' modelling takes on an added dimension because the plain English rules and membership functions can be altered or added to assess the potential impacts on the system. Once established, it may then be possible to go one step further and build dynamic spatial interaction models that can adapt to changing conditions. Performance on calibration data is ultimately far less important than how far the models can handle change.

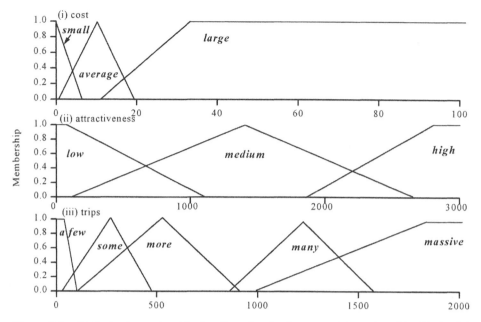

Figure 9.1 Optimized membership functions for the fuzzy model produced by Method 4.

9.4 SOME EMPIRICAL EXPERIMENTS IN LINEAR REGRESSION

Data from the 1991 census have been used to build simple fuzzy models of unemployment for Leeds. A stepwise linear regression model and results from a similar exercise carried out using a genetic programming approach (Turton *et al.*, 1997) are used as benchmarks for performance.

9.4.1 Fuzzy model formulation

The following linguistic variables were used in the fuzzy models: (a) percentage aged 16–29, (b) percentage pensionable+, (c) percentage Indian, Bangladeshi, Pakistani, (d) percentage with no car and (e) percentage unemployed at enumeration district level. These variables were selected by the genetic programming approach that yielded the best performing model. Although a wider range of variables was available, little further variation was explained when incorporated into the linear regression models. The same five methods as employed above were used to develop fuzzy models with a slight variation in Method 2. For the four input variables, the C4.5 induction software found 37 rules which were then used to produce three membership functions for variables (a) and (c) and five for (b) and (d). Percentage unemployed was partitioned into six output variables. In both Methods 2a and 2b, the Wang–Mendel algorithm was used to find the initial rulebases; however, in Method 2a, initial membership function partitioning was based on the statistical distribution of the data, while Method 2b used the membership functions from Method 1 as inputs. Method 3 then used the rulebase from Method 2b with a bit of common sense to produce a modified rulebase. A GA was then used to fine-tune each model as above. Methods 4 and 5 are the same as described previously.

Table 9.3 Performance of the fuzzy unemployment models.

Method No.	Fuzzy Model Building Techniques	Mean Sums of Squares
1	C4.5 rule induction	156.9
	after fine tuning	4.0
2a	Wang–Mendel (3 fuzzy sets)	26.6
	after fine tuning	4.0
2b	C4.5 + Wang–Mendel	8.8
	after fine tuning	4.1
3	C4.5 + Wang–Mendel (2b) + knowledge	5.8
	after fine tuning	3.9
4	restricted total optimization	5.0
5	unrestricted total optimization	5.3

9.4.2 Model goodness of fit

The stepwise linear regression model, using only the four variables listed above, yielded a mean sums of squared error of 5.0; utilizing other measures of ethnicity and affluence marginally improved the performance with a resulting error of 4.6. The genetic programming approach bred an equation with an error of 3.6. The fuzzy model results using model building Methods 1 to 5 are listed in Table 9.3.

The results show that the optimized fuzzy models in Methods 1 to 3 are able to approximate the function slightly better than the linear regression models, but with fewer variables. The best results were obtained by the Wang–Mendel algorithm (Methods 2a and b, and 3) and, once again, surprisingly good results were found by modifying the rulebase produced by the Wang–Mendel method with common sense in Method 3. This seems to indicate that, if the input and output variables are partitioned more or less correctly with a sufficient number of membership functions, the model will perform relatively well even before optimization with a GA. If a pure fuzzy model with fuzzy input and output variables is used for qualitative hypothesis testing, for example, this finding implies that neither the data nor the fuzzy model need to be extremely accurate. The GA further improves the performance of the models by moving around the membership functions and adding rule weights, but it has the disadvantage of starting from a random state, i.e. it cannot use knowledge about the membership function partitioning determined by Methods 1 and 2a and b. Further experimentation with alternative search algorithms that can use the predefined membership functions as a starting point needs to be undertaken. Methods 4 and 5 produced results equal to or slightly worse than the linear regression models. The GA found a sub-optimal solution that may have been caused by the rather large number of parameters under simultaneous optimization or premature convergence due to the mechanics of the particular GA used.

Although there were some marginal gains in performance with the fuzzy models and more so with the genetic programming approach, the relationship is essentially linear and the linear regression model is most likely an optimal performing model. Thus there was probably little room for improvement using alternative modelling techniques (Turton *et al.*, 1997). However, since fuzzy models provide an interpretable rulebase and definitions of the concepts through the membership functions, one can take a closer look into

Table 9.4 Generalized rulebase for the fuzzy unemployment model generated with Method 3

%16–29	%pensionable+	%Ind, Bang, Pak	%No Car	%Unemployed
Low to Medium ↓ Low to High ↓ Medium to High	Low to High ↓ Low to MedLow	Low (with Low) %16–29) to High (as %16–29 increases) ↓	Low to Medium ↓ Medium to High	Low ↓ Medium ↓ Higher

the behaviour of the model. Table 9.4 lists a generalized rulebase generated from Method 3 and Figure 9.2 is the corresponding set of optimized membership functions. The rulebase has been generalized because there were 70 rules generated by the Wang–Mendel method. The optimized rule weights eliminated 5 of the rules and the remaining were collapsed into a final set of 44 rules. Only the broadest patterns were extracted and placed in Table 9.4. Each column of the table corresponds to a linguistic variable and the rows represents the rules. For example, the rule in the first row of the complete rulebase was simply,

IF %16–29 is *low* AND %pensionable+ is *low* AND % Ind, Bang, Pak is *low* OR *medium* AND % with no car is *low* THEN %unemployed is *low*.

By looking at the overall trend of linguistic labels for each of the input variables as unemployment increases, one can note some of the general patterns that emerge. For instance, *low* unemployment never seems to occur when %16–29 is *high* or when % with no car is *medhigh* or *high*. Likewise, *high* and *somewhat higher* unemployment never seem to occur when %16–29 is *low* or when %with no car is *medlow* or *low*, so both of these variables are positively correlated with unemployment. As one moves from low to higher unemployment, one can see the gradual increase in the corresponding linguistic values of %16–29 and %with no car. Thus, a younger population and a lower measure of affluence occur with higher unemployment. %pensionable+ has less marked behaviour at lower unemployment, but never takes on values of *medhigh* and above at higher levels of unemployment. This may simply be a reflection of the fact that %16–29 and %pensionable+ are not completely independent variables, or that other factors such as %no car are more important variables in explaining the observed variation. The %Ind, Bang, Pak seems to have a more varied pattern with unemployment, although higher values do tend to occur with larger values of %16–29, a possible reflection of a younger age structure amongst ethnic populations. Other more specific patterns can be found by taking a closer look at the rulebase. The rules, therefore, provide a much clearer picture of the system behaviour in plain English than emerges from either the linear regression model or the GP approach. These models can then be used in a predictive capacity as they tend to have good generalization capabilities to unseen data, or the rules and membership functions can be modified to reflect the situation in other regions more accurately, allowing for a greater, potential understanding of regional differences in unemployment. Alternatively, we could extrapolate to more complex, highly non-linear spatial relationships for which linear regression models fail to capture the system behaviour and where the knowledge base is incomplete.

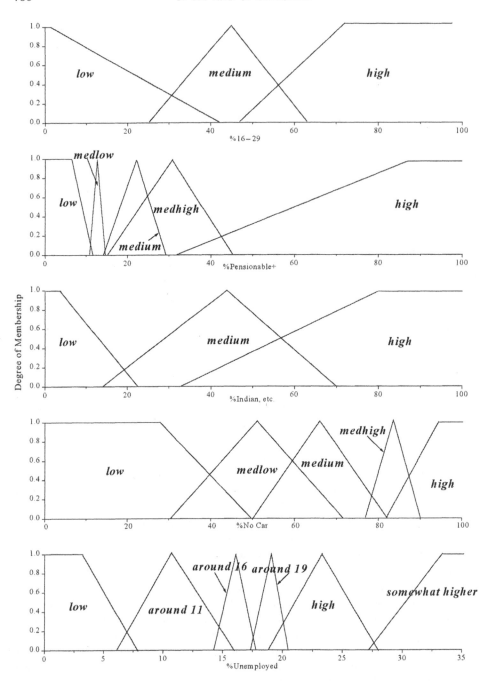

Figure 9.2   Final optimized membership functions for the fuzzy unemployment model.

9.5   CONCLUSIONS

The paper describes two examples of fuzzy models, spatial interaction and linear regression, developed by a combination of model-building techniques which can utilize expert knowledge and common sense in the fine-tuning process, or which leave the entire model-building exercise to an inductive approach. The fuzzy models of spatial interaction and unemployment were shown to achieve satisfactory performance relative to the bench-mark results, but they have the added advantage of remaining highly transparent to interpretation, so that further understanding about system behaviour can be gained. Further experimentation is still needed to determine the best combination of fuzzy-model build-ing techniques that can exploit existing knowledge as fully as possible, although promis-ing results were achieved using simple-rule discovery techniques together with common sense before optimization. It is quite apparent that fuzzy modelling has considerable potential for GIS and spatial-decision support systems, especially as the complexity of the system increases and new types of geographical model are required, which can integrate a wider range of noisy and less precise input variables. When further developed, this approach may well become an extremely powerful modelling tool that will increase the intelligent quotient of current GIS and spatial decision-support systems.

Acknowledgements

All of the census data referred to in this paper are Crown Copyright, supplied through the ESRC/JISC Census Programme. LS acknowledges support from the Crowther–Clarke Scholarship.

References

BIRKIN, M., CLARKE, G., CLARKE, M. and WILSON, A. (1996) *Intelligent GIS*, Cambridge: GeoInformation International.

BURROUGH, P.A. and FRANK, A.U. (1995) Concepts and paradigms in spatial information: are current geographical information systems truly generic? *Int. J. GIS*, **9**, 101–116.

CORDÓN, O. and HERRERA, F. (1995) A general study on genetic fuzzy systems, in G. Winter, J. Périaux, M. Galán and P. Cuesta (Eds), *Genetic Algorithms in Engineering and Computer Science*, 33–57, New York: John Wiley.

DIPLOCK, G.J. (1996) Building new spatial interaction models using GP and a supercomputer, in *Proc. 1st Int. Conf. on Geocomputation*, 213–225, Leeds: University of Leeds.

JANG, J.-S.R., SUN, C.-T. and MIZUTANI, E. (1997) *Neuro-Fuzzy and Soft Computing*, New Jersey: Prentice-Hall.

KOSKO, B. (1997) *Fuzzy Engineering*, New Jersey: Prentice-Hall.

OPENSHAW, S. (1976) An empirical study of some spatial interaction models, *Environment and Planning A*, **8**, 23–41.

OPENSHAW, S. (1996) Neural network and fuzzy logic models of spatial interaction, in *Proc. 1st Int. Conf. on Geocomputation*, Leeds: University of Leeds.

OPENSHAW, S. (1997) Building fuzzy spatial interaction models, in M.M. Fischer and A. Getis (Eds), *Recent Developments in Spatial Analysis*, Berlin: Springer-Verlag.

QUINLAN, J.R. (1993) *C4.5 Programs for Machine Learning*, San Mateo: Morgan Kauffman.

SEE, L. and OPENSHAW, S. (1997) An introduction to the fuzzy logic modelling of spatial interaction, in *Proc. 3rd Joint Eur. Conf. on Geographical Information*, 809–818, Vienna: IOS Press.

TURTON, I., OPENSHAW, S. and DIPLOCK, G.J. (1997) A genetic programming approach to building new spatial models relevant to GIS, in Z. Kemp (Ed.), *Innovations in GIS 4*, London: Taylor & Francis.

WANG, L.X. and MENDEL, J.M. (1992) Generating fuzzy rules from numerical data with applica-tions, *IEEE Transactions on Systems, Man and Cybernetics*, **SMC-22**, 1414–1427.

ZADEH, L. (1965) Fuzzy Sets, *Information and Control*, **8**, 338–353.

CHAPTER TEN

Environmental planning using spatial agents

ARMANDA RODRIGUES, CÉDRIC GRUEAU,
JONATHAN RAPER AND NUNO NEVES

10.1 INTRODUCTION

The use of agents in research has emerged from developments in Artificial Intelligence through the need to define proactive structures that require intelligent behaviour. These structures can range from very simple reactive entities, which can be globally intelligent, to complex algorithms, which use learning techniques to become more efficient in solving problems autonomously. Moreover, the application of agents into commercial applications is now becoming noticed as interface agents and search agents for the Internet became available.

The dynamic characteristics of agent systems offer the opportunity of creating new types of applications that can integrate several approaches in application development. Agents represent a new philosophy of development, where applications are built through dynamic connections between modules that are, themselves, dynamic. The idea of using agents in the manipulation of spatial information appears, therefore, natural, as it involves integrating large amounts of data with tools that implement complex spatial processes.

This paper begins by presenting agents as they have been defined in their initial areas of research. It then moves on to defining spatial agents and their current research agenda. Finally, it discusses the use of a multi-agent architecture to implement a decision model for environmental planning.

10.2 AGENTS

According to Maes (1994), an *agent* is a system that tries to fulfil a set of goals in a complex, dynamic environment: it can sense the environment through its sensors and act upon it using its actuators. The Agent paradigm has been the subject of Artificial Intelligence (AI) research for some time, but it is only now that it is beginning to spread into information technology and commercial markets. The concept of agent evolved from two fundamental threads of research:

- **Distributed Artificial Intelligence (DAI).** A branch of AI that deals with the solution of complex problems using networks of autonomous, co-operating, computational processes called agents (Adler and Cottman, 1989). These agents use explicit symbolic representations of their environment and other agents in their problem-solving and are currently called cognitive agents;

- **Information processing in Nature (Ferrand, 1996).** This thread supports that intelligence emerges from the interactions of multiple entities, acting on direct reaction to stimuli. It is related to the connectionist paradigm, supported by the idea of building extremely structured objects or processes from colonies of very simple animals (Minar et al., 1996).

10.3 SPATIAL AGENTS

The concept of spatial agent has been introduced as a specialization of the agent concept that can reason over representations of space (Rodrigues and Raper, 1997). Spatial agents understand space as either physical or non-physical phenomena which have a spatial mapping. Current work into what can be called spatial agents can be classified in three specific areas (Rodrigues *et al.*, 1997): spatial simulation, spatial decision-making and interface agents in GIS. These will be described below.

10.3.1 Spatial simulation

The implementation of spatial simulation is largely connected to the use of Cellular Automata (Wolfram, 1994). The idea of simple components that, as a whole, produce complicated patterns of behaviour can be compared to sets of reactive agents, interacting simple entities, acting on reaction to stimuli (Ferrand, 96). This thread is already being pursued at the Santa Fe Institute, where the swarm simulation system was developed with the objective of providing researchers with standardized simulation tools (Minar *et al.*, 1996). A Swarm is a collection of independent agents interacting through discrete events. Swarms can be used to build truly structured dynamic hierarchical systems, as each agent in a Swarm may be a complex system (another Swarm) and vice versa. Swarms have already been used to model settlement patterns using the individual choices of local agents that interact with each other and with the environment (Dibble, 1996). In the field of archaeology, agents representing prehistoric households were developed to understand how changes in the environment might have led to the creation of compact villages at specific points in time, as opposed to dispersed hamlets when conditions were different.

Ferrand (1995) quotes work at LAMA-IGA (France) where the limitations of Cellular Automata for dealing with complex diffusion processes led to the conversion of the work towards Multi-Reactive-Agents Systems (MRAS). According to Ferrand, these systems can be built on raster or vector data. When working with raster data, the raster will represent the environment, while agents are external processors that act upon it. In the case of vector data, the notion of environment disappears and it is necessary to define which objects belong to the environment and which will be active structures represented as agents.

10.3.2   Spatial decision-making

In the area of multi-criterion spatial decision support, Ferrand (1996) proposes the use of agents to solve the complex spatial optimization problems encountered in the search for least environmental impact area for infrastructures and to support and simulate the exchange and dynamics of spatial representations and policies. SMAALA, a Support System for Spatial Analysis, developed at IMAG in Grenoble, integrates sensitivity maps given by experts, the structural constraints of the project and political positions through the use of reactive agents that represent objects in space. Each political position is provided by a different representation of the area of study, and the process of negotiation is then simulated through the use of attraction/repulsion forces that represent the constraints presented by each actor in the process. If the system manages to reach an equilibrium, the final planning decision will have been reached.

In Papadias and Egenhofer (1995), agents are being used in qualitative collaborative planning. Agents represent topological, direction and distance constraints that are applied to a spatial planning problem. The process of planning is executed in three steps: modelling the various constraints in a unified framework (enabling interoperability among agents); checking the satisfiability of the imposed constraints; and searching the spatial database using these constraints. The focus is on using spatial access methods that can efficiently process qualitative constraints (represented by agents).

10.3.3   Interface agents in GIS

Interface Agents are semi-intelligent, semi-autonomous systems that assist users when dealing with one or more computer applications (Kozierok and Maes, 1993; Maes, 1994). The metaphor used is that of a personal assistant collaborating with the user in their work environment. Current GIS applications, manipulating large volumes of data in complex and intense processes, have been considered difficult to use and lacking in flexibility. The use of interface agents in GIS has been shown to improve the usability of applications.

Campos *et al.* (1996) describes a knowledge-based interface agent for ARC/INFO that receives and processes users' requests in plain English. The agent takes this information and generates sequences of commands that ARC/INFO can understand. If the concepts known to the user are not confirmed by the ARC/INFO database, it interacts with the user to clarify the misconception. After execution the agent delivers and presents the results to the user.

Another experiment in the field of GIS interface agents was an Interface Agent Architecture for the drawing and plotting tool of the Smallworld GIS (Rodrigues and Raper, 1997). The Smallworld GIS is an Object-Oriented (OO) GIS which was developed in an OO language called Magik. This language is part of the Smallworld system, and most of the code that forms the GIS is available for update by the user. It is, therefore, very easy and natural to include monitoring structures and add additional features to the code. Because this agent architecture was not meant to use any reasoning capabilities, Magik proved sufficient for its implementation. These agents not only give suggestions to the user on how to create drawings and plots, but also perform or finish tasks when enough information has been supplied to them.

10.4 SPATIAL AGENTS FOR AN ENVIRONMENTAL PLANNING DECISION SUPPORT SYSTEM

Addressing the problems related to the inflexibility and low performance presented by GIS packages today, the necessity for new methodologies in GIS-related software development has become apparent. The use of agents implies the implementation of applications as the dynamic integration of several modules that can autonomously manage the information they are responsible for.

In this section we will describe the MEGAAOT (Modelling Geographic Elements for Environmental Analysis in Land-use Management) project. This project was identified as a good example for an experiment in multi-agents systems because it integrates relevant areas of research described in section 10.3: Simulation of changes in the characteristics of elements (resulting in the generation of relevant information for decision-making in environmental planning), and the creation of autonomously-managed applications that will extend the interface provided by the chosen GIS package.

10.4.1 MEGAAOT

The MEGAAOT project, initially conceived for application at the municipal level and implemented on GIS generic platforms, defines intelligent connections between a GIS database, analytical operators and environmental models to create environmental, inter-related scenarios. The importance of environmental quality components for each land use (geographical elements of the areas studied) is estimated by a system of weights (called evaluation perspectives) and the relations between these elements are modelled according to the type of generated effects, the components' sensitivity to these effects and their space diffusion. The conceived decision-support model can be considered as an evolving, dynamic method for planning activities in location-allocation problems (Grueau and Rodrigues, 1997). The maintenance of several evaluation perspectives aims to provide alternative options for planning actions that can be analyzed and discussed at several levels. Changes in the characteristics of the elements produce simulations of the propagation of environmental risk, oriented towards selected effects and levels of sensitivity to environmental quality components.

The existence of such a type of system provides a standard approach for planning evaluation methodologies and a new basis for normative procedures. Three evaluation levels have been defined in the model: the expert, the municipal and the public. The expert level requires the intervention of a team of planning experts to define a system of dependencies between the model components and a set of rules for normalization in the definition of weights. At the municipal level a team of technicians defines the set of weights that implement their municipality policy and perspective, following the normalization rules defined at the previous level. The public can demonstrate its preoccupations (Shiffer, 1992) by suggesting modifications to the perspective applied by the municipality. The model provides a higher level of transparency often lacking in the decision process by allowing non-technical users to interact with its implementation, modify its criteria and evaluate the results of these changes.

The project was implemented as a spatial decision support system for environmental and municipal planning focusing on the possibility of simulating the effects of human actions and land use transformations on an interactive basis. The integration of several simulations will provide a ground for the quantitative measure of the application of simulated planning decisions.

Conceptual definition

The decision model was based on the following components:

- **Sensitivities (S).** The environmental quality components were classified and weighed against the geographical elements structure producing the concept of environmental sensitivity (biodiversity, quality of superficial water, air quality, soil quality, acoustic quality and landscape quality).

- **Effects (E).** The actions resulting from human activity which are capable of decreasing the environmental quality of the studied area. In this project the following effects were considered relevant: water release, habitat destruction, solid residue release, noise emission, air emission and erosion.

- **Propagation scenarios (CP).** A map of the potential spatial diffusion of an effect resulting from a land use transformation. In this project, propagation can be effected through superficial water, air, underground water or landscape. The propagation scenarios are calculated from physical elements of space and have been implemented using cartographic modelling processes.

- **Attenuation scenarios (A).** An attenuation scenario represents the attenuating potential of each geographical element when related with one type of effect.

The sensitivities, effects and attenuation scenarios are evaluated expertly through a system of weights that qualifies them for each geographical element. The propagation scenarios are calculated from physical elements of space.

Results

The components described above are combined to generate the following intermediate and final results:

- **Simulation lines (L_{ikjl}).** The impact of one effect is calculated by combining the effect's value map with the associated sensitivity map, the chosen propagation scenario and the relevant attenuation values. The result is called a simulation line representing the potential environmental risk for the current set of land use parcels in one defined moment t. The functional representation of the simulation can be expressed in the following way:

$$L_{ikjl}(GE_t) = S_i(GE_t) \ \theta \ E_k(GE_t) \ \theta \ CP_j \ \theta \ A_l(GE_t) \tag{10.1}$$

Where GE_t is the set of geographical elements representative of one moment t; θ is the function that enables the combination of two simulation components (in this case grid multiplication). $S_i(GE_t)$, $E_k(GE_t)$ and $A_l(GE_t)$ represent the mapping of, respectively, one of the sensitivities, effects and attenuation scenarios associated with the current set of geographical elements; CP_j is the representation of the chosen propagation scenario for this simulation.

- **Environmental performance and risk.** The potential environmental performance of the area results from the generated simulation line and enables the assessment of development of the geographic space. Environmental performance is represented as a

$2^1/_2$ D metaphorical model (Figure 10.2) to increase visual perception. The environmental risk is obtained by inverting the values of environmental performance.

■ **Decision criteria.** One modification in geographical elements between time t and $t + 1$ generates two simulation lines. The impact of this change can be measured as a variation in volume which can be calculated in area, volume and depth.

■ **Integration.** The definition of integration rules enables the estimation of a general situation or one oriented towards one or several of the defined components. It is possible to evaluate the results from simulations based on one specific theme or on a combination of themes. For example, the simulation in one specific propagation scenario may be oriented towards all the components of environmental quality, or, alternatively, one effect may be evaluated in all propagation scenarios.

Initial implementation

The first implementation of the model, although successful in the results presented by its application in the field, resulted in a heavy and redundant computational GIS application. The connections between components, created statically at development time, did not consider the evolution of data and processes. This resulted in low performance and lack of flexibility. The main problems identified were:

■ The transitive evolution of the system, resulting from changes in evaluation criteria, land use or physical space, was not reflected in the implementation. It was difficult to include new connections identified and to manage or remove old ones becoming out-of-date.

■ Every sequence of changes had to be accounted for as an algorithm.

■ The generation of simulation steps, although implemented as an automatic process, was static and of extensive implementation.

■ The information produced at several states of the planning process and/or according to different planning approaches was difficult to manage.

Defining a classic computational algorithm to implement the model is not a complete solution, for the information as well as the way it is linked may evolve during execution. The constraints and possibilities given by the information can vary according to changes in the area under study.

We propose that the use of agents, given their properties of autonomy and adaptability to the environment, is the best solution for making this model effective. Furthermore, it is our belief that the potential of the model, being much larger than its current performance, can be enlarged through a multi-agent approach.

10.4.2 MA-MEGAAOT

Multi-agent MEGAAOT implements the MEGAAOT Decision Support Model using agent technology. The connections between the decision components are established at run-time and reviewed with changes in the application environment. These connections are dynamically implemented and reviewed through the maintenance of ordered graphs representing the flow of information evolution in the system.

Agent architecture

The architecture designed to implement and control the data generated by the model includes the following agent components (Figure 10.1):

- ***Geographical elements agent*.** The agent that controls the defined geographical elements. It monitor updates and structural changes in the information related to those elements. It takes the system through the evolution determined by the changes and interacts with the components' agents to spread the evolution of the system.

- ***Components' agents (sensitivities, effects, propagation and attenuation agents)*.** One agent exists for each component of the model. This agent controls the information related to the component it is responsible for and interacts with the simulation and integration agents to inform them of changes in the information that they have used as input.

- ***Simulation agents*.** For each new simulation, one tool agent is created. This agent will then control the type of information used to create that tool and re-execute it whenever that information changes (for example, to create new moments of the simulation); simulation agents communicate with integration agents to inform them on changes in simulation data or on the need to integrate new states of the system.

- ***Integration agents*.** These agents are similar to simulation agents; they differ in the type of process they represent and control, which involves the manipulation of simulation information.

- ***Decision agent*.** This agent uses the information generated by integration agents to enable best-location spatial decisions.

The interactions between agents (presented in Figure 10.1), although currently defined for municipal applications, are internally conceived as dynamic parameters of the system. These interactions are part of an ordered graph that can be adapted to new situations as the system evolves.

Agent Architecure

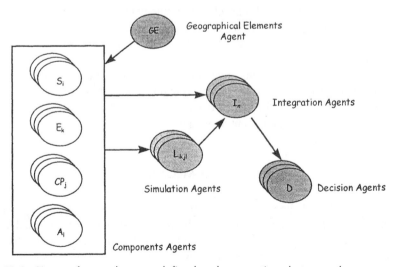

Figure 10.1 Types of control agents defined and connections between them.

System development

The Agent system is under development at the Portuguese Centre for Geographic Information (CNIG). There are still possibilities for improvement but there are already relevant statements to make. Specifically, the created spatial agents provide:

■ **System dynamic connections.** If one spatial component or criteria changes during execution, this change will be reflected in all the components that depend from the former. Therefore, all these components must be updated. This operation is activated autonomously by the spatial agent responsible for the changed component. The system of dependencies is provided by a dynamic rule base.

■ **Agent-based generation of simulation lines.** The combination of all the decision components generates the possible simulation lines (Figure 10.2). This process is automated through the use of a rule base representing the possibilities for combination (as several components are incompatible). The existence of this set of rules enables compatibilities to change during run-time and new components to be included. The generation of new simulation lines depends on the specific components agents and on compatibility information between them.

■ **Maintenance of several states and perspectives of study.** Changes in land use will represent different time-states of study. Although requiring the generation of information dependent from land use, it may be necessary to keep the previous states as well as compare changes in data and simulations (Figure 10.2). The implementation of components' agents automates the creation and maintenance of state-based information.

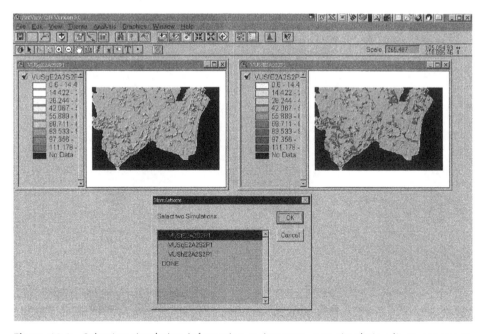

Figure 10.2 Selecting simulation information to integrate: two simulation lines, representing environmental risk at two different moments in time, are integrated in order to generate decision criteria.

■ **Generation of propagation scenarios (spatial propagation diffusion).** Propagation
information may prove difficult to handle according to the different components that
may become relevant during run-time. The generation of propagation scenarios may
be agent-enabled at the local element level to reflect global diffusion. This decision is
dependent on the types of elements available and on the complexity of the diffusion
processes.

Decision-making in MA-MEGAOOT

The creation of the multi-agent structure has already shown some results, enabling the
extension of the model with the possibility of generating information on the best location
for land-use change and the optimization of the resulting decision parameters. The study
of several simulation lines, combining the water release effect propagated in superficial
water and oriented towards sensitivity to water quality, resulted in a knowledge base of
integration results. These results compare several states that implement the same land-use
change at different locations. These are then used to generate estimated results to elements not yet used as sources of simulation. The performance of each non-tested element
is simulated according to its land use, area and surrounding elements and the chosen
decision is performed in order to confirm the decision. The result of the simulation is
added to the knowledge base, adding knowledge to the system that generated it.

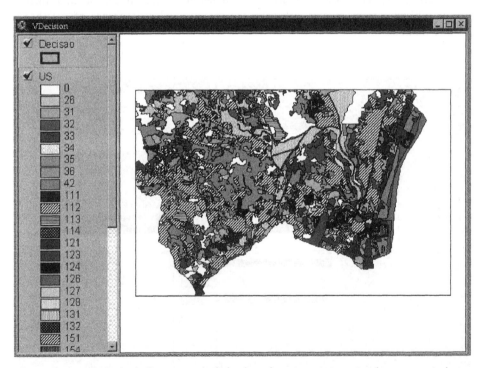

Figure 10.3 Black thick lines surround the best location sites to implement an industry
land-use change of a specific size. The perspective was one of water release effect regarding
the sensitivity of water quality.

The results presented have been generated using information from previous integration
results.

Potential improvements in decision-making

The decision-making process is currently under refinement. The simulation and integration results have generated profiles of the types of element which represent best choices for specific types of change-implementation. This has led to the realization that one knowledge base of decision results would become too complex to handle, according to different simulation and integration perspectives and concerned with the possible changes to implement. Accordingly, an embedded spatial agents system is being created to handle the complexity of this part of the system. The decision agent analyzes each element as potential for the implementation of the specific type of needed change, according to their physical properties and the planning normative of the area. Parcel agents for the elements that have not been ruled out will be created, and these agents will rank themselves according to their possibilities of performance in the decision model. Each agent will use previous results of the model, as well as information from adjacent agents, to evaluate their fitness as locations for the specific change to be implemented. The aim is to have agents deciding if the area they represent is sufficiently fit to be used.

10.5 CONCLUSIONS

It is our belief that the use of an agent architecture in MEGAAOT has enhanced the performance and the potential of its decision model. The dynamic and proactive structure of the application facilitates the automatic creation and maintenance of the necessary information for simulation and integration. The information thus generated enables the preview of further results and decision-making based on generalization. The system generates a suggestion of best decision which can then be confirmed by the generation of the actual simulation. The result of the simulation is integrated with the system's knowledge, enabling it to learn in the case of wrong suggestion. Spatial agents that can evaluate themselves as parts of a solution will simplify the complexity of decision-making, with unfit agents stepping out of the selection process.

Acknowledgements

The research performed by Armanda Rodrigues is funded by grant PRAXIS XXI/BD/ 2920/94 of the PRAXIS XXI/JNICT Programme. The basis for the MEGAAOT project evolved from research related with the 1ST/CNIG (Portugal) sigla project funded by JNICT, Portugal.

References

ADLER, R.M. and COTTMAN, B.H. (1989) A Development Framework for Distributed Artificial Intelligence, in *Proc. 5th Conf. on Artificial Intelligence Applications*, IEEE.

BAEIJS, C., DEMAZEAU, Y. and ALVARES, L. (1996) SIGMA: Application of Multi-Agent Systems to Cartographic Generalization, in *Proceedings of The Seventh European Workshop on Modeling Autonomous Agents in a Multi-Agents World*, Eindhoven, January 1996.

CAMPOS, D., NAUMOV, A. and SHAPIRO, S. (1996) Building and Interface Agent for ARC/INFO, in *Proc. 1996 ESRI User Conf.*, Palm Springs, *Cal., 20–24 May, 1996*, <http://www.esri.com/ resources/userconf/proc96/TO50/PAP049/P49.HTM>.

DIBBLE, C. (1996) Theory in a Complex World: Agent-Based Simulations of Geographic Systems, in *Proc. 1st Int. Conf. on Geocomputation*, Sept. 1996, Leeds, UK.

FERRAND, N. (1995) Multi-Reactive-Agents Paradigm for Spatial Modelling, in *Proc. GISDATA Workshop on Spatial Modelling*, June 1995, Stockholm.

FERRAND, N. (1996) Modelling and Supporting Multi-Actor Spatial Planning Using Multi-Agents Systems, in *Proc. 3rd NCGIA Conf. on GIS and Environmental Modelling*, January 1996, Santa Fe, USA.

GRUEAU, C. and RODRIGUES, A. (1997) Simulation Tools for Transparent Decision Making in Environmental Planning, in *Proc. 2nd Annual Conf. on GeoComputation*, 26–29 August, 1997, University of Otago, Dunedin, New Zealand.

KOZIEROK, R. and MAES, P. (1993) A learning interface agent for scheduling meetings, in *Proc. ACM-SIGCHI Int. Workshop on Intelligent User Interfaces*, Jan. 1993, Florida, USA.

MAES, P. (1994) Modeling Adaptive Autonomous Agents, *Artificial Life*, **1**, 135–162.

MINAR, N., BURKHART, R., LANGTON, C. and ASKENAZI, M. (1996) *The Swarm Simulation System: a Toolkit for Building Multi-Agent Simulations*, June 1996, <http://www.santafe.edu/projects/swarm/overview.ps>.

PAPADIAS, D. and EGENHOFER, M. (1995) Qualitative Collaborative Planning in Geographical Space: Some Computational Issues, *NCGIA Initiative 17 Position Paper, 16–19 Sept., 1995*, <http://www.ncgia.ucsb.edu:80/research/i17/htmlpapers/papadias/Papadias.html>.

RODRIGUES, A., GRUEAU, C., RAPER, J. and NEVES, N. (1997) Research on Spatial Agents, in *Proc. 3rd Joint Eur. Conf. on Geographic Information (JEC-GI '97)*, 16–18 April, 1997, Vienna.

RODRIGUES, A. and RAPER, J. (1997) Defining Spatial Agents, in *Spatial Multimedia and Virtual Reality Research Monograph*, to be published, London: Taylor and Francis.

SHIFFER, M.J. (1992) Towards a Collaborative Planning System, Environment and Planning B: Planning and Design, **19**, 709–722.

WOLFRAM, S. (1994) *Two-Dimensional Cellular Automata, Cellular Automata and Complexity: Collected Papers*, Reading, Massachussets: Addison-Wesley.

CHAPTER ELEVEN

Fuzzy geodemographics: a contribution from fuzzy clustering methods

ZHIQIANG FENG AND ROBIN FLOWERDEW

11.1 INTRODUCTION

Geodemographics, as a successful application of geographical information systems (GIS), enables marketers to predict behavioural responses of consumers based on statistical models of identity and residential location (Goss, 1995). In North America and Europe geodemographics has become a rapidly growing information industry (Flowerdew and Goldstein, 1989; Watts, 1994). At present geodemographics is not only applied in the private sector but in public issues such as deprivation analysis, fundraising for charity and healthcare analysis.

Nevertheless, there are some weaknesses which have drawn criticism from academic and commercial sectors. Among the weaknesses are updating problems, the ecological fallacy, and labelling problems (Flowerdew, 1991). Openshaw (1989a) proposed there are at least four major sources of errors and uncertainties in all the available systems. How much discrimination is provided by the geodemographic systems is a question regarding the efficiency of current geodemographic systems. In evaluating direct mail a success rate of the order of one per cent is generally viewed as quite acceptable (Openshaw, 1989a). Geodemographics is viewed as a revolution in contemporary marketing research and practice. Still on the subject of direct mailing, a one per cent response means 99 per cent failure. Thus, the question is: can we do better?

11.2 FUZZY GEODEMOGRAPHICS

As early as 1989 Openshaw suggested that the concept of fuzziness could usefully be employed in geodemographics. Fuzziness refers to uncertainties arising from imprecision and ambiguity. Usually there are two kinds of fuzziness existing in geodemographics. The first kind is fuzziness in attribute space. The neighbourhood classification is Boolean so that it ignores the fact that many enumeration districts (EDs) may be quite close in the taxonomic space to more than one neighbourhood type; an ED can be misclassified to a wrong group according to the algorithm. Areas may differ by only very small amounts in the classification but be assigned to very different clusters.

The second kind is geographical fuzziness which consists of two aspects. One refers to the problem of linking postcodes to EDs, the other to genuine neighbourhood effects. The linkage problem arises from different causes, and results in the error that incorrect postcodes are included in a cluster and correct postcodes are excluded (Openshaw, 1989a, 1989b). The neighbourhood effect is the geographical phenomenon that people who live near to each other tend to share or simulate some behavioural characteristics despite other differences (Openshaw et al., 1994). This is just the first principle of geodemographics. In geodemographic classifications these effects have been implicitly exploited at ED level. However, boundaries of EDs are not natural in terms of partitioning the demographic characteristics and economic behaviour of residents (Morphet, 1993). It is reasonable to assume that people who live in neighbouring EDs still possess similar characteristics to some degree, marginal or minor, or aspire to the same lifestyle as their neighbours. Genuine neighbourhood effects can be taken as the effects beyond those denoted at ED scale. Neighbouring EDs to a target ED should be handled to take genuine neighbourhood effects into account. At present, these neighbourhood effects are completely missed. Openshaw (1989a, 1989b) stressed that

> in many applications it is these fuzzy areas, largely missed by conventional geodemographic targeting, that hold the greatest promise for new prospects and business. Therefore the real challenge is how to make best use of this fuzziness factor within geodemographics.

A fuzzy geodemographic system needs to be built to tackle the classification fuzziness and genuine neighborhood effect issues. Two questions are typical. Suppose cluster 15 is a target cluster for mailing. Where are the areas near to cluster 15 in classification space? If an ED is in cluster 15, which EDs in the neighbouring area should be taken as near to it due to the genuine neighbourhood effect? These fuzzy queries are listed as fuzzy spatial targeting among several spatial analysis techniques useful for geodemographics (Openshaw, 1995a).

Openshaw (1989a, 1989b) suggested a solution to handle fuzziness and to set up a fuzzy targeting system. The objective is to allow for simultaneous variation in the fuzziness of both the geo- and the demo-parts of geodemographics. By dividing the levels of both sources of uncertainty into a small number of classes, and then classifying a mailing response by both variables using a training data set, an optimal subset of areas is selected with a degree of fuzziness to provide the best results for targeting customers for a special task. The neural nets classifier can also be applied to the classification, and fuzziness of the results is preserved in a particular easy-to-use form (Openshaw et al., 1994).

11.3 FUZZY CLUSTERING

Suppose that we have N areas and we want to classify them into k clusters. Each area is identified by n attributes including demographic, ethnic, housing, household composition, social-economic, health and travel-to-work.

A fuzzy geodemographic area is defined as:

$$(X_i, \mu_{ij}) \quad i = 1, \ldots, N, j = 1, \ldots, k; \quad \sum_j^k \mu_{ij} = 1 \tag{11.1}$$

X_i denotes an area, μ_{ij} denotes its membership in cluster type j. If μ_{ij} is 1 then X_i falls precisely in the cluster type j; if μ_{ij} is 0 then X_i is definitely not in cluster type j. Areas may belong partially to several clusters at the same time with μ_{ij} varying from 0 to 1.

The n attributes of an area can be represented by a vector

$$(x_{i1}, x_{i2}, \ldots x_{in}) \tag{11.2}$$

Suppose we can define a representative vector for each cluster and view the distance between an area and the representative vector as the distance between the area and the cluster. The representative vector of cluster j is of the form:

$$(\lambda_{j1}, \lambda_{j2}, \ldots \lambda_{jn}) \quad j = 1, 2, \ldots, k \tag{11.3}$$

The Euclidean distance between X_i and λ_j:

$$d_E(X_i, \lambda_j) = \sqrt{\sum_p^n (x_{ip} - \lambda_{jp})^2} \tag{11.4}$$

The membership of X_i to cluster j can be defined as:

$$\mu_{ij} = \frac{d_E^{-1}(X_i, \lambda_j)}{\sum_j^k d_E^{-1}(X_i, \lambda_j)} \tag{11.5}$$

More generally the membership value is defined as:

$$\mu_{ij} = \frac{(d_E(X_i, \lambda_j))^{-2/(m-1)}}{\sum_j^k (d_E(X_i, \lambda_j))^{-2/(m-1)}} \tag{11.6}$$

where m is called the fuzzy exponent.

This is the formula defined in the method of fuzzy-k-means (FKM) or fuzzy-c-means (FCM). Fuzzy clustering can be viewed as an extension of traditional clustering. In crisp clustering a sample must be forced into one of several clusters according to some rule or algorithm. However, in fuzzy clustering a sample could belong to different groups at the same time by assigning degrees of memberships to different groups. This also means that fuzzy clustering can reduce the distortion produced by outliers or intermediate cases to a minimum, especially for poorly separated data. Thus the classification results cope with the data structure more closely.

In fact FKM is an extension of conventional k-means classification. The widely used algorithm of FKM was developed by Dunn (1974) and Bezdek (1981). FKM uses an objective function as the base of the algorithm. By iteratively minimizing the objective function, the final results are reached. The objective function is defined as (Bezdek, 1981; Bezdek *et al.*, 1984):

$$J_m(U, \lambda) = \sum_i^N \sum_j^k \mu_{ij}^m d^2(x_i, \lambda_j) \tag{11.7}$$

The exponent m determines the degree of fuzziness of the solution. When m declines towards the smallest value of 1, the solution is close to a crisp partition. As m increases, the degree of fuzziness will grow till $\mu_{ij} = 1/k$ for every sample, when no clusters are found. There is no theoretical or computational evidence distinguishing an optimal m (Bezdek *et al.*, 1984). The optimal value comes from experiments using real data.

Fuzzy clustering has been used widely in various disciplines such as ecology (Equihua, 1990), geochemistry (Frapporti *et al.*, 1993), soil science (McBratney and de Gruijter, 1992) and regionalization (Harris *et al.*, 1993).

11.4 BRIEF OUTLINE OF THE LANCASHIRE DATA

1991 census data for Lancashire is selected for our research. There are 2 971 enumeration districts in the county excluding shipping and restricted EDs. Our main concern here is to test the efficiency of the fuzzy clustering method and we decide the classification is to be general purpose. We simply choose the same 44 variables as listed in Openshaw and Wymer's paper (1995). These variables include demographic, housing, economic and social attributes. We use values per 10 000 in order to reduce information loss due to small figures.

The summary statistics for Lancashire data show that the majority of the 44 variables are positively skewed. However there are some variables which are negatively skewed, or normally distributed. A few attributes have the same values at 20, 40, 60 percentiles which are equal to their median values. To avoid the bias that variables with large variance have more effect on the classification, the 0–1 standardizing method is employed to the whole data set. Figure 11.1 (see colour section) is a map of Lancashire with ED boundaries, and the Lancaster area is indicated by a rectangle box which will be used to demonstrate the following results.

11.5 FKM RESULTS AND THEIR APPLICATIONS

First we select the fuzzy exponent m separately as 1.25, 1.5, 1.75 and 2 and the number of clusters from 4 to 20. After processing we find the results for m equal to 1.5, 1.75 and 2 are very fuzzy since membership values of different clusters are very close, which makes them very difficult to interpret. We discard these results, therefore, and pick m equal to 1.25 for further analysis. Bear in mind that we are only trying to test the fuzzy clustering method in setting up a geodemographic system; we simply use the results of five clusters (groups) as an example to show how it works.

We use Euclidean distance in the FKM which gives equal weights to all the attributes. According to the characteristics of the demographic, housing, and social variables, Group A could be described as struggling, B as climbing, C as established, D as aspiring and E as thriving.

In order to make further analysis and to clarify the results of the fuzzy classification, defuzzification is needed to transform the membership values into a representation of the hardened groups. A sample belongs to a group if it has the highest membership in that group. The intermediate type is identified if an ED has approximately the same membership values on two, or more than two, groups which sum to one. There are 131 intermediates most of which are transitional types between two neighbouring groups. Hardened groups are presented in Figure 11.1. Group E is distributed most widely in the county, especially in rural areas. Group B boasts the largest number of EDs among the five groups. Groups, A, B, C, and D usually occur in urban and suburban areas.

Table 11.1 Customer profile for product X: base-GB.

Group type	Profile count	Profile percentage	Base percentage	Index
A	150	7.5	5.1	147
C	1 040	52.0	14.1	369
D	750	37.5	10.0	375
F	60	3.0	17.0	18
Totals	2 000	100	46.2	

Table 11.2 Customer profile for product X detailed by fuzzy membership value.

Group type	Group subtype	Profile count	Profile percentage	Base percentage	Index
A	0.5–0.749(A1)	38	1.9	3.0	63
	0.75–1.0(A2)	112	5.6	2.1	267
C	0.5–0.749(C1)	400	20	5.9	339
	0.75–1.0(C2)	640	32	8.2	390
D	0.5–0.749(D1)	190	9.5	3.9	244
	0.75–1.0(D2)	560	28	6.1	459
Totals		1 940	97	29.2	

The membership values are stored in an INFO database. The fuzzy classification map of Lancaster uses proportional symbols to represent the membership values at each ED (Figure 11.2–see colour section). From the map we can visualize the constitution of membership values at each ED and the variation over the whole area.

Geodemographic profiling can be undertaken in more detail by relating not only the groups but the different degrees of membership to them. The first step will be the same as that in the conventional method, and only for those groups which have possible prospects do we need to analyze the cross-reference between customer profile and different membership degrees in the groups.

We use a fictional example from Sleight (1993) to make a demonstration. Table 11.1 is assumed to be a layout of profiles provided by a geodemographics agency. It is a customer profile for a fictional product. In fuzzy set theory a core subset is often defined as a representative crisp set of the fuzzy set. The membership value 0.75 is usually determined as a threshold of the core (Wang, 1994). We then can answer a query such as 'Where are typical areas for class A?' The query is reduced to the question, 'Where are the areas which have membership values greater than and equal to 0.75 on group A?' This division is useful when a detailed profiling is needed to search for potential customers. Below we use 0.5 to 0.749 and 0.75 to 1.0 of membership values to partition hardened groups into two sub-groups.

In the results of FKM the group types for customer profiling are hardened types. The profile count is a count of the customers in each of the group types. The profile percentage is the percentage of total customers represented by each group type. The base percentage is the percentage of the base area represented by each group type. The index is equal to profile percentage divided by base percentage multiplied by 100. Groups A, C, D are selected due to their high index values. We can set up another table with detailed customer profiles (Table 11.2).

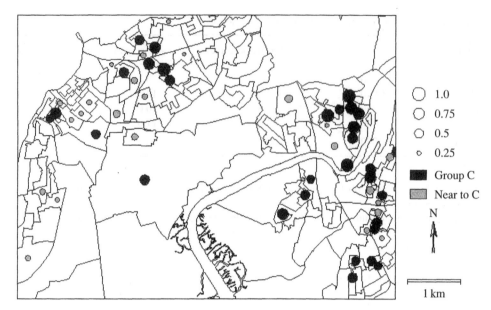

Figure 11.3 Membership values for Group C in the Lancaster area.

The index for subset A1 of group A increases to 267, larger than that of subset D1. If we recommend the order of groups for direct mailing for the client, then it should be D2, C, A2, D1 and A1. In addition, because of the high index values for C1 and C2 it may be necessary to calculate the index of subgroups which have membership values smaller than 0.5. If these subgroups have high index values, they are also prospects for targeting. The detailed customer profile provides further possibility for target marketing which allows users more benefits for less expense.

Even if the index of subgroups of group C with membership value less than 0.5 is not high, areas similar to group C should be identified because of fuzziness in attribute space. Thus the problem becomes identifying the neighbourhoods which have similar characteristics to class C. This is the first question fuzzy geodemographic systems are intended to answer. If we have used hardened groups in customer profiling, then the threshold is 0.5 – i.e. any case with larger than 0.5 membership value on some group is attributed to that group. Therefore membership value near to the group can be defined as greater than and equal to 0.25. The question is then reduced to finding out the areas which have membership values greater than and equal to 0.25 on group C. The map in Figure 11.3 shows areas which are in and near to group C in the Lancaster area.

Genuine neighbourhood effects can also be handled on the basis of fuzzy results. Neighbourhood effects reflect spatial interaction between contiguous areas. The membership values of groups represent the degree of similarity for each group. After profiling with customer survey data, membership values are actually taken as a degree of similarity to a certain type of consumption behaviour. When the effect of geographical space is considered, the membership values of each ED must be adjusted.

The adjustment is up to the corresponding membership values of neighbouring EDs and their spatial relations. We design a mathematical method to recalculate the membership values of EDs. The new membership value for group k of one ED is:

$$\mu_i' = \alpha\mu_i + \beta\frac{1}{A}\sum_j^n w_{ij}\mu_j \qquad (11.8)$$

where μ' and μ are respectively the new and old membership of an ED, and n is the number of contiguous EDs to the ED. α, β are respectively weights to old membership and the mean of membership values of surrounding EDs and they observe:

$$\alpha + \beta = 1 \qquad (11.9)$$

The proportion of α to β reflects the relative importance of demographic characteristics and spatial interaction. If $\alpha = \beta = \frac{1}{2}$, this implies that spatial interaction has the same influence as demographic features on consumer behaviour. $\alpha > \beta$ implies that the customer behaviour is mainly determined by an area's demographic attributes. w_{ij} is the weight measuring the amount of interaction between a pair of EDs. The weight is decided by distance between ED centroids or the length of common boundary between EDs, or both. We follow what Cliff and Ord (1973) used in defining the weight to calculate spatial autocorrelation statistics:

$$w_{ij} = d_{ij}^{-a} p_{ij}^b \qquad (11.10)$$

where d_{ij} refers to distance between the study ED and the jth contiguous ED, p_{ij} is the length of common boundary between them and a, b are parameters. Positive values of a, b give greater weights to pairs of EDs which have shorter distances between their centres, and which have long common boundaries. In addition, w_{ij} needs to be scaled so that the average of weighted membership values is in the range of 0 to 1. Thus A is used as such a parameter.

In our research we use a simplified version of the formula to calculate the new membership. The new membership value is taken as the average of the old membership value and the mean membership value of the neighbouring EDs weighted by distances between EDs as a decay measure.

There are respectively 58.5 and 40 per cent of EDs which increase and decrease membership values. Among them, 16.1 per cent decrease by more than 0.1 and 18.1 per cent increase by more than 0.1. Figure 11.4 is the map for the Lancaster area showing new membership values on group C. In the southwest corner several EDs which are near to group C in Figure 11.3 now disappear, because they are surrounded by low-membership EDs. In the east a few EDs which do not occur in Figure 11.3 now turn up as 'near to group C' because neighbouring EDs are in group C. These new membership values are a combination of measures for both demographic and geographical attributes of an area. These values can then be used to target areas which are near to the prospect group. In a word, by using the adjusted membership values the second question in fuzzy spatial targeting can be treated as the first one. The EDs near to the prospective group can be identified as those with membership values greater than or equal to 0.25.

11.6 DISCUSSION AND CONCLUSION

Based on fuzzy set theory, FKM is a fuzzy clustering method which is very useful for detecting patterns in a poorly-separated dataset. We use FKM to classify Lancashire 1991 census data into five groups, and thus to set up a fuzzy geodemographic system.

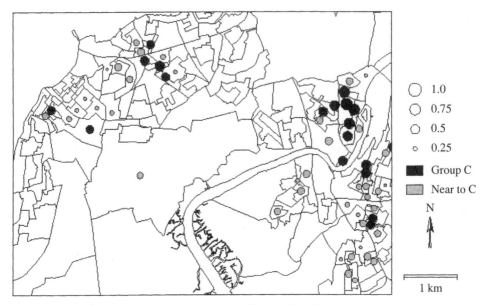

Figure 11.4 Adjusted membership values for Group C in the Lancaster area.

The large amount of information ignored by traditional methods is stored as the membership values and can be used to conduct fuzzy queries. The membership values from FKM can be employed to undertake the customer profiling more precisely. The membership values of EDs in each group present information about the EDs with respect to the centre values of each group. Through these membership values we can defuzzify the results into hardened groups and intermediates. By combining the hardened groups and degrees of membership we can understand the variation both within groups and between groups. The variation within group is ignored in the conventional method. We can search for areas which are not in a group type but similar to it in the classification space. A method is proposed to take genuine neighbourhood effects into account. Adjusted membership values are extracted using membership values of contiguous EDs and measures of spatial relationship. These new membership values contain information both from demographic and geographic space. Thus genuine neighbourhood effects are incorporated into the membership values which are then used in the customer targeting.

Fuzzy clustering is a mathematical method based on distance in attribute space. The resulting membership values are easy to understand, interpret, and use in further study. The neighbourhood effect is successfully tackled by interpreting the memberships as degrees of similarity to consumer behaviour. Although there is a solution using distance measures from hard clustering to handle fuzziness in attribute space, the fuzzy clustering method is much more straightforward and the resulting membership values are subject to further analysis.

GIS are more than simple 'delivery systems' in geodemographic systems. The geo-part of geodemographic systems is the most important characteristic for targeting customers. Therefore, in the GIS era geographical knowledge and GIS tools together with new methodologies such as fuzzy logic can and should do more to improve the efficiency of geodemographic systems.

Acknowledgement

Thanks are due to Professor Anthony C. Gatrell who provided valuable comments.

References

BEZDEK, C.J. (1981) *Pattern Recognition with Fuzzy Objective Function Algorithms*, London: Plenum Press.

BEZDEK, C.J., EHRLICH, R. and FULL, W. (1984) FCM: the fuzzy c-means clustering algorithm, *Computer and Geosciences*, **10**, 191–203.

CLIFF, A.D. and ORD, J.K. (1973) *Spatial Autocorrelation*, London: Pion.

DUNN, J.C. (1974) A fuzzy relative of the ISODATA process and its use in detecting compact, well separated clusters, *J. of Cybernetics*, **3**, 32–57.

EQUIHUA, M. (1990) Fuzzy clustering of ecological data, *J. of Ecology*, **78**, 519–534.

FLOWERDEW, R. (1991) Classified residential area profiles and beyond, *Research Report No. 18*, NWRRL, Lancaster University.

FLOWERDEW, R. and GOLDSTEIN, W. (1989) Geodemographics in practice: Developments in North America, *Environment and Planning A*, **21**, 605–616.

FRAPPORTI, G. and VRIEND, S.P. (1993) Hydrogeochemistry of the shallow Dutch groundwater: interpretation of the national groundwater quality monitoring network, *Water Resource Research*, **29**, 2993–3004.

GOSS, J. (1995) 'We know who you are and we know where you live': the instrumental rationality of geodemographics systems, *Economic Geography*, **71**, 171–198.

HARRIS, T.R., STODDARD, S.W. and BEZDEK, J.C. (1993) Application of fuzzy set clustering for regional typologies, *Growth and Change*, **24**, 155–165.

McBRATNEY, A.B. and DE GRUIJTER, J.J. (1992) A continuum approach to soil classification by modified fuzzy k-means with extragrades, *Journal of Soil Science*, **43**, 159–175.

MORPHET, C. (1993) The mapping of small-area census data – a consideration of the effects of enumeration district boundaries, *Environment and Planning A*, **25**, 1267–1277.

OPENSHAW, S. (1989a) Making geodemographics more sophisticated, *J. Market Research Society*, **31**, 111–131.

OPENSHAW, S. (1989b) Learning to live with errors in spatial databases, in M. Goodchild and S. Gopal (Eds), *The Accuracy of Spatial Databases*, London: Taylor and Francis.

OPENSHAW, S. (1995a) Marketing spatial analysis: a review of prospects and technologies relevant to marketing, in P. Longley and G. Clarke (Eds.), *GIS for Business and Service Planning*, Cambridge: Geo Information International.

OPENSHAW, S. and WYMER, C. (1995) Classifying and regionalizing census data, in S. Openshaw (Ed.) *Census User's Handbook*, Cambridge: Geo Information International.

OPENSHAW, S., BLAKE, M. and WYMER, C. (1994) Using neurocomputing methods to classify Britain's residential areas, *Proc. GISRUK '94*, 11–13 April, Leicester, UK.

SLEIGHT, P. (1993) *Targeting Customers*, Henley-on-Thames: NTC Pubs.

WANG, F. (1994) Towards a natural language user interface: an approach of fuzzy query, *Int. J. Geographical Information Systems*, **8**, 143–162.

WATTS, P. (1994) European geodemographics on the up, *GIS Europe*, **3**, 28–30.

CHAPTER TWELVE

A qualitative representation of evolving spatial entities in two-dimensional topological spaces

CHRISTOPHE CLARAMUNT, MARIUS THÉRIAULT
AND CHRISTINE PARENT

12.1 INTRODUCTION

The search for a better understanding of natural and anthropic phenomena is one of the major objectives of natural and human sciences. A wide range of applications in both the sciences and planning disciplines have developed models to explain and simulate real-world evolution. The validation and efficiency of these models are highly dependent on data quality and consistency in both space and time. This has led to a need for an integrated representation of geographical and historical data which corresponds as closely as possible to the way the real world changes. Although various conceptual models combining space and time have been proposed (Langran, 1992; Cheylan, 1993; Peuquet, 1994; Frank, 1994; Worboys, 1994; Galton, 1995), research on temporal GISs (TGIS) is still in an early stage of development. The development of TGISs requires a formal foundation comparable to the work realized for spatial models. A comprehensive framework and mathematical models of spatio-temporal relationships are needed for the development of flexible and efficient geo-historical information systems.

The model we propose in this paper addresses the evolution process of a single spatial entity and the multi-linear evolution of a set of spatial entities. The evolution of an independent spatial entity involves changes in its size and shape as well as movement processes. The multilinear evolution represents changes in a set of spatial entities that are of the same spatial type and that cover the same area; the identified processes are the split of a spatial entity and the unification or reallocation of several spatial entities. The framework is based on a conceptual representation of spatio-temporal processes defined in previous work (Claramunt and Thériault, 1995, 1996). In this paper, we extend the former model semantics with a qualitative representation based on a mathematical description of spatio-temporal processes. This development aims at providing a sound model support (i.e. physically and logically possible spatio-temporal relationships). This spatio-temporal model may later contribute to support TGIS software development (e.g. spatio-temporal query languages and search algorithms). The resulting rules may be used to validate spatio-temporal relationships if they are explicitly stored in the database, or otherwise to compute and discover them.

The paper is organized as follows. We present a set of basic properties of space in Section 2, and a set of minimal topological relationships in Section 3. Properties of time are examined in Section 4. Section 5 introduces the notion of change and a set of temporal properties that link evolving entities. Section 6 describes the fundamental properties of evolving spatial entities. In Section 7 we develop a qualitative model for representing the linear evolution of a spatial entity. In Section 8 the multi-linear evolution of spatial entities is modelled. Section 9 concludes the paper and outlines future work.

12.2 PROPERTIES OF SPACE

Two types of phenomena can be represented in space. The first type is physically bounded and is usually assimilated, for scientific purposes, to a semi-closed system (e.g. a building or a car). There is little uncertainty about the limits of these real-world entities and their semantic definitions are generally universal and unambiguous. The second type is continuously spread over space with no obvious limits but with transition areas between states measured on continuous scales. For purposes of analysis, scientists often define interpreted semantic classes that cover persisting regions of space sharing similar values (Newton-Smith, 1980). The land they cover is delimited using a specific set of taxonomic rules that may change according to the application. Resulting territorial extensions are defined using geometric entities (e.g. soil categories, air masses, administrative regions). Even if it is not always straightforward to distinguish between real and interpreted entities, their distinction allows for the identification of particular properties: a real entity is represented in a topological space by a mapping function between a frame of reference and the topological space (e.g. a car or a building projected in a 2-dimensional space), while an interpreted entity is directly represented in the topological space.

The core of the proposed model is based on spatial entity primitives which describe real or interpreted entities that are discrete and located in space. A class of spatial entities represents a collection of spatial entities sharing similar generic characteristics. Each of these classes is defined by a type with its own generic and specific attributes. A spatial attribute describes the geometry of a spatial entity. For the purpose of this paper, its value domain is restricted to 2-dimensional (2-D) space, consisting of: points (0 dimension), lines (1 dimension, non-closed without self-intersection) and polygons (2-dimensional contiguous patch of space with no hole). The properties of topological spaces have been used extensively as a mathematical support for GIS design. They provide a formal support for the expression of spatial relationships (Alexandroff, 1961). Let us consider a 2-D topological space with a minimum set of properties: a spatial relationship model based on point-set topology (Herrings, 1991) and a metric based on Euclidean geometry. Topological operators are defined in topological spaces, informally:

- A spatial entity (e) is a subset of the topological space (X) that is described by a complement, a closure, an interior and a boundary.
- The complement of a set $(X - e)$ is the collection of points that surrounds the set.
- The closure of a set (e^-) is the intersection of all the surrounded sets.
- The interior of a set (e°) is the collection of points completely surrounded by the set.

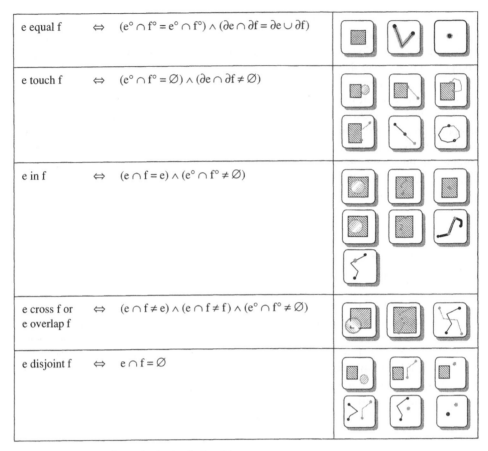

e equal f	$\Leftrightarrow$	$(e^\circ \cap f^\circ = e^\circ \cap f^\circ) \wedge (\partial e \cap \partial f = \partial e \cup \partial f)$
e touch f	$\Leftrightarrow$	$(e^\circ \cap f^\circ = \varnothing) \wedge (\partial e \cap \partial f \neq \varnothing)$
e in f	$\Leftrightarrow$	$(e \cap f = e) \wedge (e^\circ \cap f^\circ \neq \varnothing)$
e cross f or e overlap f	$\Leftrightarrow$	$(e \cap f \neq e) \wedge (e \cap f \neq f) \wedge (e^\circ \cap f^\circ \neq \varnothing)$
e disjoint f	$\Leftrightarrow$	$e \cap f = \varnothing$

Figure 12.1 Spatial topological relationships.

■ The boundary of a set (∂e) is the set of points that intersects both the set and its complement. The boundary of a point feature (p) is empty ($\partial p = \varnothing$). The boundary of a non-connected line is the set of the two separate end-points.

12.3 TOPOLOGICAL RELATIONSHIPS IN SPACE

Spatial relationships in 2-D spaces are formally defined by topological operators (Egenhofer, 1991; Cui, 1993; Clementini, 1993; Abdelmoty, 1995). They generally differ in the type of spatial entities they cover (e.g. region to region, line to region relationships). Clementini proposes a generic model based on the topological intersections of two entities in space. This model is nearly independent of the geometric type (i.e. point, line or polygon) and presents the advantage of identifying a reduced number of relationships. Those topological relationships are summarized in Figure 12.1. A spatial entity (e or f) is defined as a 2-dimensional point-set (polygon, line or point) where its three components (interior, boundary and closure) are connected.

We group the cross and overlap relationships, since they only differ by the geometric dimension of the resulting intersection and this distinction is not significant for the purpose of this paper. The touch and cross/overlap relationships are not defined for points.

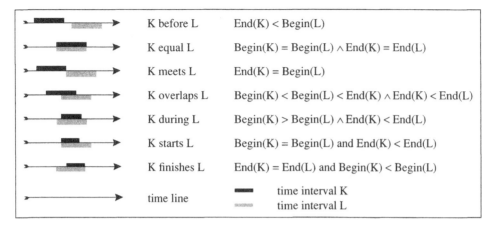

	K before L	End(K) < Begin(L)
	K equal L	Begin(K) = Begin(L) ∧ End(K) = End(L)
	K meets L	End(K) = Begin(L)
	K overlaps L	Begin(K) < Begin(L) < End(K) ∧ End(K) < End(L)
	K during L	Begin(K) > Begin(L) ∧ End(K) < End(L)
	K starts L	Begin(K) = Begin(L) and End(K) < End(L)
	K finishes L	End(K) = End(L) and Begin(K) < Begin(L)
	time line	time interval K time interval L

Figure 12.2 Temporal relationships.

With the exception of the 'in' operator, all the other relationships are symmetric. The 'in' and 'equal' relationships are transitive.

12.4 BASIC PROPERTIES OF TIME

For the purpose of the following model, time defines a structure to label and link evolving spatial entities (things in time). This temporal structure is represented as a linear time-line which supports partial ordering. The characterization of time is defined with metric properties (measured time leading to chronology) and topological properties (ordered events and interaction networks). A temporal metric assigns values to things in time and measures duration, whereas a temporal topology allows the study of properties which are preserved under all continuous transformations.

Two time structures are generally used to represent a temporal property: point-oriented (instant) or interval-oriented logics (duration). Although our model uses an interval approach, it is also applicable, with minor adaptations, to an instant logic approach. Moreover, it is not dependent on the choice of a continuous or discrete time-line representation. A continuous time-line means that any interval of time is divisible since it is expressed with real numbers; whereas a discrete time-line means a finite representation where time is measured using integers. We use Allen's interval logic as it provides a complete set of temporal operators (Allen, 1984). Temporal operators are defined through mutually exclusive relationships between time intervals (Figure 12.2). Begin(I) and End(I) are temporal operators which provide respectively the beginning and ending instants of a time interval I.

Change happens when there is an alteration of any attribute describing a real-world entity or a geographic region (Newton-Smith, 1980). However, spatio-temporal models must distinguish between trend and fluctuation changes (Miles, 1992). A trend is an average change over a significant period of time leading from one state (the value of any attribute) to another (Bornkamm, 1992). In structural terms, a trend changes the equilibrium point of the entity sub-system (i.e. entity attribute values) for a significant duration, without modifying the entity type (as opposed to the mutation process). Conversely, a fluctuation is a rapid change in attributes that is not stable within the application's time granularity, or that is related to observation bias. Fluctuations are reversible and oscillate around the average trend values. Therefore, the proposed model will only consider trend variations as they lead to more meaningful changes.

12.5 FROM CHRONOLOGY TO HISTORY

To give a more precise signification of change, we define the state of an entity:

- The state of a spatial entity e, denoted as e_i, represents a stable value (without trend change) of this entity within an interval of time I_i as (Allen, 1984): Holds(e_i, I_i) $\Leftrightarrow$ $\forall\ t \in I_i$, Holds_at(e_i, t).

- An entity state holds throughout every sub-interval included in its own temporal interval: Holds(e_i, I_i) $\wedge J \subseteq I_i \Rightarrow$ Holds(e_i, J).

We characterize the evolution of an entity (e) with its two successive linear states (e_i and e_{i+1}) defined respectively within I_i and I_{i+1} intervals of time. In order to represent the linear structure of time, these intervals are constrained:

Holds(e_i, I_i) $\wedge$ Holds (e_{i+1}, I_{i+1}) $\Rightarrow$ [End(I_i) $\leq$ Begin(I_{i+1})].

Evolving spatial entities use an interaction network to link their two successive ($e_i, \ldots, h_m$) and ($e_{i+1}, \ldots, h_{m+1}$) state sets defined respectively within ($I_i, \ldots, I_m$) and ($I_{i+1}, \ldots, I_{m+1}$) intervals of time (Figure 12.3). Their respective time intervals are also constrained:

[Holds(e_i, I_i), ..., Holds(h_m, I_m)] $\wedge$ [Holds(e_{i+1}, I_{i+1}), ..., Holds(h_{m+1}, I_{m+1})] $\Rightarrow$ [Maximum (End(I_i), ..., End(I_m)) $\leq$ Minimum(Begin(I_{i+1}), ..., Begin(I_{m+1}))].

Linear and multi-linear evolving entities share temporal properties (Figure 12.4):

- e_i and e_{i+1} are two immediate states of an entity if and only if End(I_i) = Begin(I_{i+1}).

- e_i and e_{i+1} are two delayed states of an entity if and only if End(I_i) < Begin(I_{i+1}).

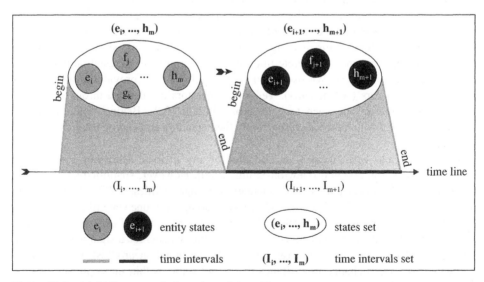

Figure 12.3 Multi-linear evolution of spatial entities.

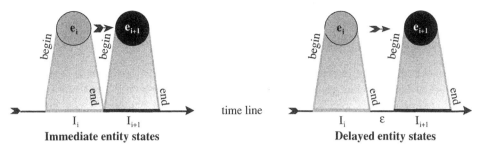

Figure 12.4 Successive entity states.

- $(e_i, \ldots, h_m)$ and $(e_{i+1}, \ldots, h_{m+1})$ are two immediate sets of entity states if and only if $\text{End}(e_i) = \ldots = \text{End}(h_m) = \text{Begin}(e_{i+1}) = \ldots = \text{Begin}(h_{m+1})$.

- $(e_i, \ldots, h_m)$ and $(e_{i+1}, \ldots, h_{m+1})$ are two delayed sets of entity states if and only if $\text{End}(e_i) = \ldots = \text{End}(h_m) = \text{Begin}(e_{i+1}) = \ldots = \text{Begin}(h_{m+1})$ is false.

- Delayed and immediate states represent two possible time-ordering relationships between two non-overlapping sets of time intervals.

The represented world is discrete in both space and time. Discontinuity in the life of an entity is assumed in the time line and represented by delayed entity states. This property is required in order to represent episodic phenomena (e.g. successive recorded positions of a moving spatial entity, weekly measurement of the water quality at a beach station). There is an analogy between this time discontinuity and the discrete structure of space used in physics (e.g. in the Bohr–Rutherford model where the position of an electron changes in a discrete, quantized manner).

12.6 SPACE AND TIME PROPERTIES

A space and time model completes the static properties of a spatial entity by providing dynamic characteristics and integrating the following properties for an evolving spatial entity:

- The state of a spatial entity, denoted (e_i), is characterized by its location in space and by its form, defined by the union of its interior and its boundary $(e_i^\circ \cup \partial e_i)$.

- The location of a spatial entity state, denoted $P(e_i)$, is defined by a projection of its centre of gravity into a 2-D Euclidean coordinate system $(\Re^2) \Rightarrow$

  $P \colon X \to \Re^2$ and $e_i \to \{x(e_i), y(e_i)\}$.

Two states of two distinct spatial entities e_i and e_j share the same location in space (i.e. $P(e_i) = P(e_j)$) if and only if $[x(e_i) = x(e_j)] \wedge [y(e_i) = y(e_j)]$. Galton defines the position of an entity as the total region of space it occupies at a given time (Galton, 1995). We go further and represent the total region of space occupied by a geographic entity as the union of its interior and boundary plus the location of its centre of gravity. The distinction can thus be made between contraction, expansion and deformation from translation and rotation evolution through analysis of the locational change (centre of gravity displacement) and the topological intersection between two successive states of an entity.

Describing change for GIS leads to the definition of the spatial evolution of geographic entities. A spatial evolution represents a trend change over space. It is characterized by the modification of the values of its spatial properties (geometrical and topological components).

The model integrates two kinds of evolving spatial entities:

- Those that are free to change, and whose semantics are defined from an application point of view without geometrical or topological constraints (e.g. a vehicle, a person, a hurricane). They can be modelled using a single entity approach and the usual time line.

- Those whose changes are constrained by topological, positional or structural relationships with other entities (e.g. within a cadastral database, lots must remain adjacent and cover the entire county at any time, even when they exchange their territory). They need a network time model to express their joint evolution and to retain these complementary relationships using the multi-linear evolution approach.

The distinction between independent and related entities must be fixed in accordance with the application requirements because the same entity type (e.g. vehicles) may be seen as free to move for some purpose (e.g. a ship sailing over the sea), or restricted by the geometry and topology of a transportation network for another application (e.g. a car moving within a city's street network).

12.7 LINEAR EVOLUTION OF AN INDEPENDENT SPATIAL ENTITY

This section presents the basic spatio-temporal relationships defined for an independent entity using the concept of linear evolution. Two successive immediate or delayed states (e_i and e_{i+1}) of a single spatial entity represent an isomorphic evolution if they are respectively members of two isomorphic topological spaces X_1 and X_2. This happens if and only if there exists an isomorphism H (i.e. a mapping function which maintains the shape and size of spatial entities) between the X_1 and X_2 topological spaces. Considering the positional and the morphologic properties of two successive spatial entity states (location, shape, size and orientation) we obtain the following five minimal and orthogonal cases (Figure 12.5):

- *Stability* is provided to express the invariance of spatial components.
- *Expansion and contraction* are inverse forms of homeomorphism i.e. a mapping function that preserve shape and orientation (i.e. size change, location may change).
- *Deformation* is an evolution that implies possible simultaneous change in size and orientation (i.e. shape change, size and location may change).
- *Translation* involves movement in space while maintaining shape and size (isomorphism), and orientation (i.e. location change).
- *Rotation* implies locational stability and isomorphism but change of orientation.

Isomorphic spatio-temporal relationships (rotation and translation) have an equivalence property: the topological relationships between e_i and e_j in X_1 are said to be equivalent to the topological relationships between e_{i+1} and e_{j+1} in X_2 if and only if there is an isomorphism between X_1 and X_2. Non-isomorphic transformations are restricted to possible relationships between successive states, and three cases are identified: expansion, contraction and deformation. Other combinations are not valid according to the following restrictions:

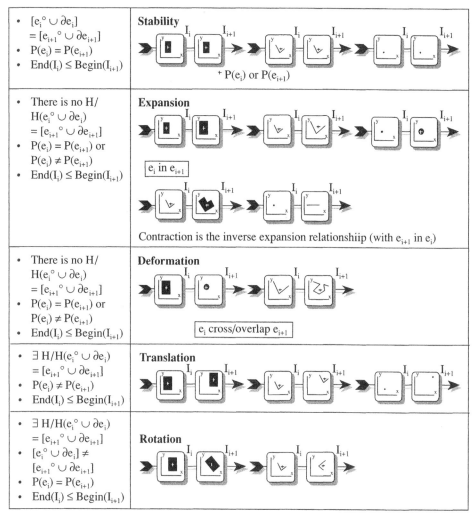

• $[e_i{}^\circ \cup \partial e_i]$ $= [e_{i+1}{}^\circ \cup \partial e_{i+1}]$ • $P(e_i) = P(e_{i+1})$ • $End(I_i) \le Begin(I_{i+1})$	**Stability**
• There is no H/ $H(e_i{}^\circ \cup \partial e_i)$ $= [e_{i+1}{}^\circ \cup \partial e_{i+1}]$ • $P(e_i) = P(e_{i+1})$ or $P(e_i) \ne P(e_{i+1})$ • $End(I_i) \le Begin(I_{i+1})$	**Expansion**
• There is no H/ $H(e_i{}^\circ \cup \partial e_i)$ $= [e_{i+1}{}^\circ \cup \partial e_{i+1}]$ • $P(e_i) = P(e_{i+1})$ or $P(e_i) \ne P(e_{i+1})$ • $End(I_i) \le Begin(I_{i+1})$	**Deformation**
• $\exists\, H/H(e_i{}^\circ \cup \partial e_i)$ $= [e_{i+1}{}^\circ \cup \partial e_{i+1}]$ • $P(e_i) \ne P(e_{i+1})$ • $End(I_i) \le Begin(I_{i+1})$	**Translation**
• $\exists\, H/H(e_i{}^\circ \cup \partial e_i)$ $= [e_{i+1}{}^\circ \cup \partial e_{i+1}]$ • $[e_i{}^\circ \cup \partial e_i] \ne$ $[e_{i+1}{}^\circ \cup \partial e_{i+1}]$ • $P(e_i) = P(e_{i+1})$ • $End(I_i) \le Begin(I_{i+1})$	**Rotation**

Figure 12.5 Spatio-temporal relationships for an evolving spatial entity.

■ *Cross and overlap* are commutative relationships: e_i cross/overlap e_{i+1} is equivalent to e_{i+1} cross/overlap e_i.

■ *Touch and disjoint* relationships are not used since they imply a *de facto* translation.

■ *Topological equality* is irrelevant as well because it is not compatible with the non-isomorphic property.

Particular restrictions apply for point evolution:

■ *Point deformation* and rotation are irrelevant since a point is dimensionless and has no shape or orientation.

■ *Point contraction* is equivalent to disappearance.

■ *Point expansion* implies a geometric type mutation from point to line or from point to polygon.

Combining these five basic operators is sufficient to model complex movements of an independent entity in a 2-D Euclidean space. However, we need networks to describe complex movements of droves grouping many independent entities (Cheylan, 1993) or to represent co-ordinated evolution of related spatial entities (Frank, 1994; Claramunt and Thériault, 1996). We discuss the multilinear evolution of spatial entities in the next section.

12.8 MULTILINEAR EVOLUTION OF RELATED SPATIAL ENTITIES

As the range of phenomena that can be processed in a temporal GIS is probably inexhaustible, modelling all possible cases of multilinear evolution involving related spatial entities is a complex task and one far beyond the scope of this paper. This section analyzes the redistribution of geometric space among a set of geographic entities of the same spatial type that collectively define an exhaustive coverage and exclusive partition of a subset of the topological space. Since multilinear relationships often imply sets of evolving spatial entities, we introduce a set of symbols to describe their characteristics and operations:

- $\{E_1\} = (e_i, f_j, \ldots, h_m)$ and $\{E_2\} = (e_{i+1}, f_{j+1}, \ldots, g_k)$ are two consecutive (immediate or delayed) sets of spatial entities.
- $\{E_1\}^\circ$ is the union of interiors of $e_i, f_j, \ldots, h_m$.
- $\partial\{E_1\}$ is the union of boundaries of $e_i, f_j, \ldots, h_m$.
- $\{E_1\}^\circ \cap \{E_2\}^\circ$ is the intersection of the interiors of $(e_i, f_j, \ldots, h_m)$ with those of $e_{i+1}, f_{j+1}, \ldots, g_k$.
- $\partial\{E_1\} \cap \partial\{E_2\}$ is the intersection of boundaries of $(e_i, f_j, \ldots, h_m)$ with those of $e_{i+1}, f_{j+1}, \ldots, g_k$.

A redistribution of space involves a set of spatial entities and their topological components (Figure 12.6). Together they make an exhaustive coverage of an overall territory (e.g. counties in a country) but they do not overlap (i.e. exclusive partition of space). This generic class of multilinear evolution has applications in various fields: property management, political division of land to avoid jurisdictional conflicts, street names management to provide unique addresses. They are also useful for building diachronic series with census data, to avoid the effects of an eventual re-zoning of statistical areas.

Figure 12.6 illustrates three basic forms of geometric redistribution used to restructure a network of connected lines or adjacent polygons: split, union and re-allocation. The multilinear evolution adds an additional dimension to these networks since they must also be connected in time. All these spatio-temporal relationships share the same basic requirements:

- Successive entity sets must have the same geometric type.
- They must exist during their specific time intervals and they must use line or polygon primitives to display their position in space.
- They must be connected in time (immediate or delayed successors).
- They entirely cover the same geometric extension (exhaustivity) without overlapping (exclusivity).

Successive entity sets are constituted with mutually exclusive geometric primitives that are connected in space (connected lines or adjacent polygons) before and after the change

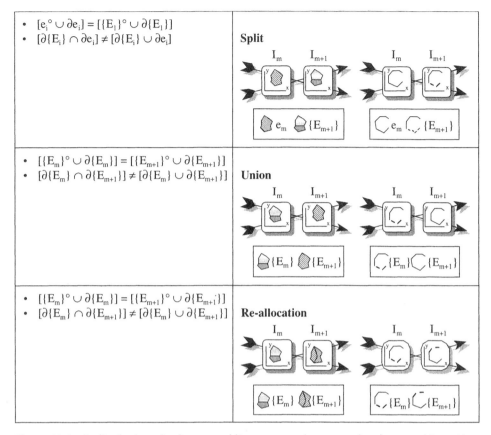

- $[e_i{}^\circ \cup \partial e_i] = [\{E_1\}^\circ \cup \partial\{E_1\}]$
- $[\partial\{E_i\} \cap \partial e_i] \neq [\partial\{E_i\} \cup \partial e_i]$

Split

- $[\{E_m\}^\circ \cup \partial\{E_m\}] = [\{E_{m+1}\}^\circ \cup \partial\{E_{m+1}\}]$
- $[\partial\{E_m\} \cap \partial\{E_{m+1}\}] \neq [\partial\{E_m\} \cup \partial\{E_{m+1}\}]$

Union

- $[\{E_m\}^\circ \cup \partial\{E_m\}] = [\{E_{m+1}\}^\circ \cup \partial\{E_{m+1}\}]$
- $[\partial\{E_m\} \cap \partial\{E_{m+1}\}] \neq [\partial\{E_m\} \cup \partial\{E_{m+1}\}]$

Re-allocation

Figure 12.6   Redistribution of polygons and line segments between related geographic entities.

occurs. There is some change in the internal boundaries layout. The only factor that differentiates them is the composition of the previous and subsequent entity sets; they are one-to-many (split), many-to-one (union) or many-to-many (re-allocation) relationships. The last is most difficult to implement because it implies an algorithm or an heuristic to find the minimum previous set of entities ($\{E_1\}$) that perfectly overlaps ($\{E_1\} \cap \{E_2\} = \{E_1\} \cup \{E_2\}$) a corresponding minimum subsequent set of entities ($\{E_2\}$).

12.9   CONCLUSIONS

The development of temporal GIS needs formal models to represent relationships in both space and time. Qualitative reasoning, which is widely used to define spatial relationships, must therefore be extended to the temporal dimension to provide a set of well-defined, homogeneous spatio-temporal relationships.

This paper introduces a general method to represent qualitative relationships for evolving spatial entities. The model allows the formalization of the evolution of spatial entities. It is flexible and applies to 2-dimensional spatial entity types. We identify a set of minimal spatial evolutions for a single spatial entity (stability, expansion/contraction, deformation, translation and rotation) and a particular case of multi-linear spatial evolutions (split, union and re-allocation). Further work concerns the model extension toward the representation of functional relationships such as replacement and diffusion processes.

References

ABDELMOTY, A.I. and EL-GERESY, B.A. (1995) A general approach to the representation of spatial relationships, Technical Report CS-95-6, Department of Computer Studies, University of Glamorgan, Pontypridd, UK.

ALEXANDROFF, P. (1961) *Elementary Concepts of Algebraic Topology*, New York: Dover Publishing.

ALLEN, J.F. (1984) Towards a general theory of actions and time, *Artificial Intelligence*, **23**, 123–154.

BORNKAMM, R. (1992) Mechanisms of succession on fallow lands, in J. Miles, W. Schmidt and E. Van der Maarel (Eds), *Temporal and Spatial Patterns of Vegetation Dynamic*, 95–101, The Netherlands: Kluwer Academic Publishers.

CHEYLAN, J.P. and LARDON, S. (1993) Toward a conceptual model for the analysis of spatio-temporal processes, in A.U. Frank and I. Campari (Eds), *Spatial Information Theory*, 158–176, Berlin: Springer-Verlag.

CLARAMUNT, C. and THÉRIAULT, M. (1995) Managing time in GIS: An event-oriented approach, in J. Clifford and A. Tuzhilin (Eds), *Recent Advances in Temporal Databases*, 23–42, Berlin: Springer-Verlag.

CLARAMUNT, C. and THÉRIAULT, M. (1996) Toward semantics for modelling spatio-temporal processes within GIS, in M.J. Kraak and M. Molenaar (Eds), *Advances in GIS Research I*, 27–43, London: Taylor & Francis.

CLEMENTINI, E., DI FELICE, P. and VAN OOSTEROM, P. (1993) A small set of topological relationships suitable for end-user interaction, in D.J. Abel and B.C. Ooi (Eds), *Advances in Spatial Databases*, 277–295, Berlin: Springer-Verlag.

CUI, Z., COHN, A.G. and RANDELL, D.A. (1993) Qualitative and topological relationships in spatial databases, in D.J. Abel and B.C. Ooi (Eds), *Advances in Spatial Databases*, 296–315, Singapore: Springer-Verlag.

EGENHOFER, M. (1991) Reasoning about binary topological relations, in O. Günther and H.-J. Schek (Eds), *Advances in Spatial Databases*, 143–160, Berlin: Springer-Verlag.

FRANK, A.U. (1994) Qualitative temporal reasoning in GIS – Ordered time scales, in *Proc. 6th Int. Symp. on Spatial Data Handling*, 410–430, London: Taylor and Francis.

GALTON, A. (1995) Towards a qualitative theory of movement, in A.U. Frank and W. Kuhn (Eds), *Spatial Information Theory: A Theoretical Basis for GIS*, 377–396, Berlin: Springer-Verlag.

HERRINGS, J.R. (1991) The mathematical modelling of spatial and non-spatial information in geographic information systems, in D. Mark and A.U. Frank (Eds), *Cognitive and Linguistic Aspects of Geographic Space*, 313–350, Nato Series.

LANGRAN, G. (1992) *Time in Geographic Information Systems*, London: Taylor & Francis.

MILES, J., SCHMIDT, W. and VAN DER MAAREL, E. (1992) *Temporal and Spatial Patterns of Vegetation Dynamics*, The Netherlands: Kluwer Academic Publishers.

NEWTON-SMITH, W.H. (1980) *The Structure of Time*, London: Routledge and Kegan Paul.

PEUQUET, D.J. (1994) It's about time: A conceptual framework for the representation of temporal dynamics in geographic information systems, *Annals of the Ass. of the American Geographers*, **84** (3), 441–461.

WORBOYS, M.F. (1994) A unified model of spatial and temporal information, *Computer Journal*, **37** (1), 26–34.

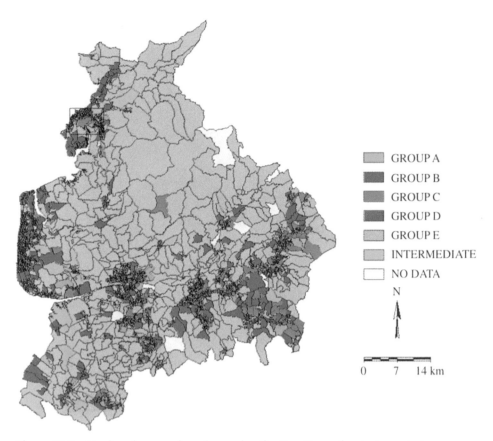

Figure 11.1 Hardened groups from fuzzy classification (Box refers to Lancaster).

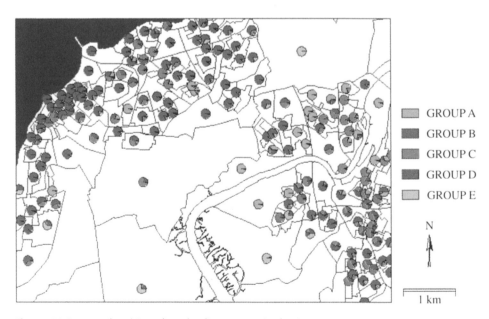

Figure 11.2 Membership values for five groups in the Lancaster area.

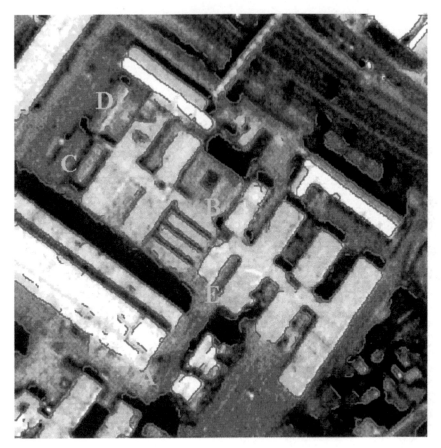

Figure 15.3 Raw image edges.

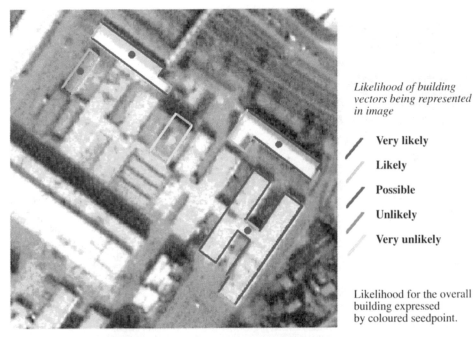

Likelihood of building vectors being represented in image

Very likely

Likely

Possible

Unlikely

Very unlikely

Likelihood for the overall building expressed by coloured seedpoint.

Figure 15.4 Buildings labelled according to the likelihood of being represented in the image.

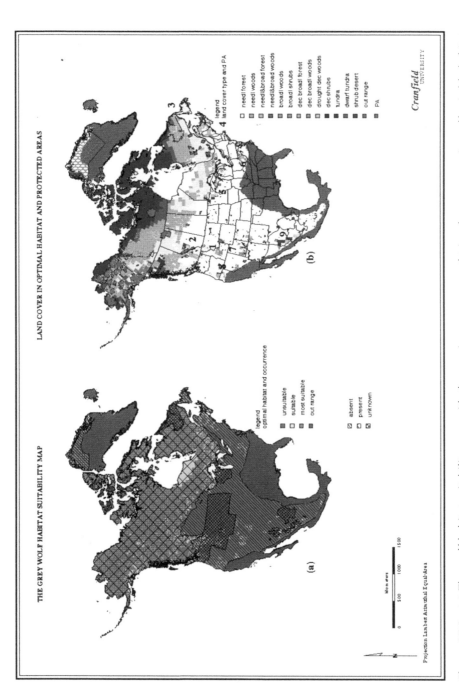

Figure 17.3 (a) The wolf-habitat suitability map with the species occurrence. (b) Land cover types in wolf-suitable habitat with Protected Areas outline (PA).

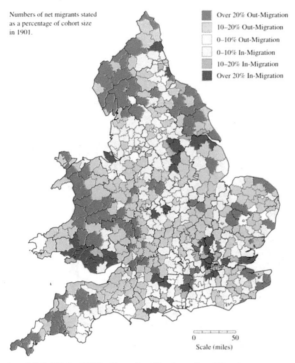

Numbers of net migrants stated
as a percentage of cohort size
in 1901.

Over 20% Out-Migration
10–20% Out-Migration
0–10% Out-Migration
0–10% In-Migration
10–20% In-Migration
Over 20% In-Migration

0   50
Scale (miles)

'Young Adults' = Aged 15–24 in 1901. Data for Registration Districts.

Figure 19.2   Net migration by young adult males, 1901–11.

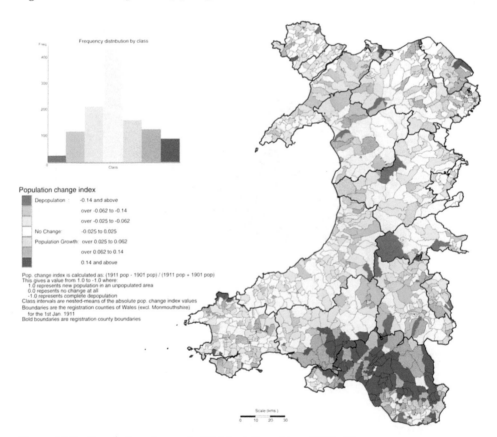

Frequency distribution by class

Population change index

Depopulation :	-0.14 and above
	over -0.062 to -0.14
	over -0.025 to -0.062
No Change:	-0.025 to 0.025
Population Growth:	over 0.025 to 0.062
	over 0.062 to 0.14
	0.14 and above

Pop. change index is calculated as: (1911 pop - 1901 pop) / (1911 pop + 1901 pop)
This gives a value from 1.0 to -1.0 where:
  1.0 represents new population in an unpopulated area
  0.0 represents no change at all
  -1.0 represents complete depopulation
Class intervals are nested-means of the absolute pop. change index values
Boundaries are the registration counties of Wales (excl. Monmouthshire)
  for the 1st Jan. 1911
Bold boundaries are registration county boundaries

Scale (kms.)
0   10   20   30

Figure 19.3   Population change in Welsh civil parishes, 1901–11.

PART FOUR

Space-time GIS

CHAPTER THIRTEEN

A fuzzy coordinate system for locational uncertainty in space and time

ALLAN BRIMICOMBE

13.1 INTRODUCTION

In a GIS context, a categorical thematic map is a topologically-ordered collection of spatial entities (points, lines, polygons, cells) in which descriptors about a theme (usually defined within a discrete classification scheme) are given spatial extent, location and boundary so that within-unit variation is expected to be less than between-unit variation, where a unit is a polygon or contiguous group of cells having a common descriptor. Thematic maps, whether in analog or digital form, have an unavoidable characteristic: uncertainty.

The term 'uncertainty' can encompass any facet of the data, its collection, storage, manipulation or its presentation as information which may raise concern, doubt or scepticism in the mind of the user as to the nature or validity of the results or intended message. It thus spans both data and model quality where the models and data manipulation encapsulate human understanding of the phenomena (Burrough et al., 1996). Errors in spatial data – that is, deviations from an accepted truth – are usually conceptualized as a contingency table in which there is a separation between the characterization of a spatial entity and its locational definition (Robinson & Frank, 1985). However, regardless of the presence of error, model and data quality may be adversely affected by carto-graphic conventions used in thematic mapping; cartographic conventions which for many have come to be *de facto* views of reality. For example, geographical boundaries are invariably portrayed as sharp lines, regardless of whether that boundary represents an abrupt change or a convenient demarcation within a trend. Hence Burrough's statement that 'many soil scientists and geographers know from field experience that carefully drawn boundaries and contour lines on maps are elegant misrepresentations of changes that are often gradual, vague or fuzzy' (Burrough, 1986). This is abetted by the capability of GIS software to infinitely zoom vectors which steadfastly remain as pixel-thin bound-aries implying much higher precision than warranted. Thus, to some extent, uncertainty in GIS is of its own making.

Testing of locational accuracy has not changed with the advent of GIS, with most map accuracy standards relying on a sample of well-defined points having no attribute ambiguity tested by reference to a survey of a higher order. Standards are usually stated in the general form: 90% of all points tested shall be correct within Xmm at map scale and are usually reported as a root mean square error. Notwithstanding that such checks are expensive and infrequently carried out, the method is not so useful in a GIS context. Not only does it result in a global figure for an entire area in which boundaries may vary quite widely in their sharpness, but well-defined points are likely to be more accurately mapped anyway, and therefore only testing them will give a biased result of data quality.

Operationally-induced degradation in quality arises through data conversion and structuring in a GIS. The results of tests vary (e.g. Maffini *et al.*, 1989; Bolstad *et al.*, 1990; Dunn *et al.*, 1990) but scale and quality of source documents and operator variability are key factors likely to result in a lowering of locational accuracy by at least ±0.125mm at map scale.

The epsilon band, ε, was introduced into GIS as a means of representing the possible positional error around a point or line (Chrisman, 1982, 1983; Blakemore, 1983). This buffer-zone approach has been widely used. Examples are its use as a fuzzy tolerance for automatic removal of spurious polygons (Zhang and Tulip, 1990) and as a mechanism of classifying mismatches in the integration of primary and secondary data (Law and Brimicombe, 1994). ε is, nevertheless, invariably applied as a global measure of point or line error without due regard to the range of boundary sharpness that may exist in reality. This range is most noticeable in polygon data layers arising out of piecemeal partitioning of near continuously varying fields in which discretization is necessary for the representation of geographical reality in a GIS. The experiments of Drummond (1987) and Edwards and Lowell (1996) using superimposed multiple photo-interpretations show this clearly. In the intervening period since the introduction of ε, 'relatively little research has been undertaken in computer cartography or GIS to improve the expressive ability of polygon boundaries' (Wang and Hall, 1996).

Recent research has focused on boundary uncertainty and recast it as being related primarily to the properties of the polygons on either side of a boundary rather than a feature of a single polygon (Wang and Hall, 1996; Edwards and Lowell, 1996). Thus a 'twain' (Edwards and Lowell, 1996) is a pair of polygons and their common boundary. Wang and Hall calculate boundary uncertainty from measures of purity of left and right polygons. Brimicombe (1997) has developed a method of recording linguistic hedges of uncertainty of polygon purity which can be used similarly to assign an uncertainty grade to common boundaries. Nevertheless, there is scope for going further and recording explicitly the *locational extent* of uncertainty occasioned by a boundary between two polygons.

13.2 FUZZY SETS IN GIS

The term 'fuzzy' has been introduced to describe uncertainty in GIS, but for the most part the term has been loosely applied to any non-binary treatment of data, particularly probabilities. Introduced by Zadeh (1965), fuzzy sets are used to handle the imprecision that characterizes much of human reasoning. The use of fuzzy-set theory proper in GIS has thus far been quite restricted (Unwin, 1995) and is reviewed in Altman (1994). One area of application has been the 'fuzzification' of data, database queries and classification schemes through the use of fuzzy membership functions, as a means of overcoming the uncertainty implicit in the binary handling of data (Wang *et al.*, 1990; Kollias and Voliotis, 1991; Burrough *et al.*, 1992). Another area of application has been to quantify verbal assessments of data quality (Gopal and Woodcock, 1994; Brimicombe, 1997). Specific to the problem of boundary sharpness, Wang and Hall (1996) have used a 'geographical boundary fuzzy set' as a global concept for a map layer and derive a membership grade to denote sharpness of boundary from the purity of the adjacent polygons. Their approach assumes that all the residual uncertainty can be assigned amongst the available classes. In reality this may not be possible. Consequently, a different approach is adopted here in which fuzzy points, lines and polygons are constructed around a fuzzy number coordinate system which gives enormous flexibility, whilst retaining intuitive simplicity.

13.3 FUZZY NUMBERS

Kaufmann and Gupta (1985) have extended the concept of fuzzy sets to arithmetic and, in doing so, define fuzzy numbers. Thus a fuzzy number is a coupling of an interval of confidence with a level of presumption. For a number A with a level of presumption α:

$$A_\alpha = [a_1^{(\alpha)}, a_2^{(\alpha)}]; \quad \alpha \in [0, 1] \tag{13.1}$$

For every level of presumption in A, there is an interval of confidence:

$$\forall_{\alpha 1, \alpha 2} \in [0, 1] : (\alpha_1 < \alpha_2) \Rightarrow ([a_1^{(\alpha_2)}, a_2^{(\alpha_2)}] \subset [a_1^{(\alpha_1)}, a_2^{(\alpha_1)}]) \tag{13.2}$$

A fuzzy number is said to be 'convex' if it has only one peak and 'normal' if that peak has $\mu_A(x) = 1$. Because of notation and computational difficulties in handling continuous fuzzy numbers, fuzzy numbers can be generalized to a triangular fuzzy number (TFN):

$$A = (a_1, a_2, a_3); \text{ where at } a_1, \mu_A(x) = 0; a_2, \mu_A(x) = 1; a_3, \mu_A(x) = 0 \text{ and } a_1 \leq a_2 \leq a_3 \tag{13.3}$$

Examples of normal convex fuzzy numbers and their equivalent TFN are given in Figure 13.1. Kaufmann and Gupta's notation of TFN (3) has been further simplified and made intuitively number-like by the author:

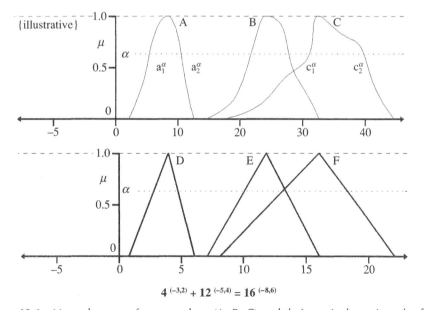

$$4^{(-3,2)} + 12^{(-5,4)} = 16^{(-8,6)}$$

Figure 13.1 Normal convex fuzzy numbers (A, B, C) and their equivalent triangular fuzzy numbers (D, E, F). The graphic TFN are representative of the example of arithmetic addition.

$$A = a_2^{(a_1, a_3)}; \text{ where at } a_1, a_3 \text{ interval of confidence } \mu_A(x) = 0 \text{ relative to } a_2, \mu_A(x) = 1$$
(13.4)

By using relative extents in the notation (instead of absolute), arithmetic computation becomes straightforward. An example of an addition of two fuzzy numbers using the author's notation is given in Figure 13.1. In addition to a full range of arithmetic operators, fuzzy numbers can also be used in Boolean operations of union (MAX) and intersection (MIN) as with fuzzy sets. Where the relative extents of a fuzzy number or TFN are equal to zero, then the number becomes a conventional crisp number. Crisp numbers are thus a special case of fuzzy number.

13.4 FUZZY COORDINATES: POINTS, LINES AND POLYGONS

Using the concept of fuzzy numbers with the author's notation, it is possible to develop and apply a fuzzy coordinate system that behaves in the same way as conventional coordinate systems, while allowing flexibility in defining the sharpness of a boundary independent of the purity of the polygon. A coordinate of easting and northing could thus be represented by triangular fuzzy numbers in the form:

$$E^{(-e, +e)}, N^{(-n, +n)}$$
(13.5)

where positive and negative signs could be taken as implicit. This concept of a fuzzy coordinate is illustrated in Figure 13.2. Although the easting and northing coordinates traditionally represent a two-dimensional space, a fuzzy coordinate is a three-dimensional concept. The third dimension concerns the level of membership $\mu_E(x)$, $\mu_N(x)$ of E, N in ordinary linguistic terms of confidence, belief, certainty or plausibility. Since most of these terms have distinct mathematical or statistical connotations and so as to avoid confusion with these terms, another term is introduced: *expectation*.

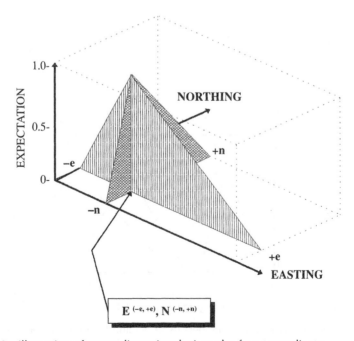

Figure 13.2 Illustration of a two-dimensional triangular fuzzy coordinate.

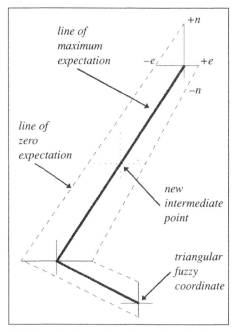

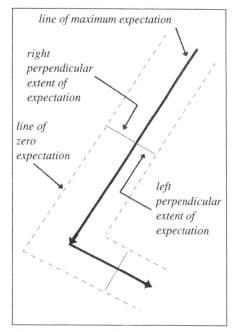

(a) triangular fuzzy lines from triangular fuzzy coordinates

(b) triangular fuzzy lines from perpendicular fuzzy extents

Figure 13.3 Triangular fuzzy lines.

The *triangular fuzzy coordinate* (TFC) given in Figure 13.2 shows that there is a point, of zero-dimension in the traditional sense, at E, N (at the intersection of the two fuzzy numbers) where confidence in both numbers is at a maximum so that $\mu_E(x) = \mu_N(x) = 1$. Maximum expectation (confidence) in a TFC is always normalized to 1. This is the *point of maximum expectation* (PME). On either side of this point and along the coordinate axes, expectation diminishes linearly to $\mu_E(x) = \mu_N(x) = 0$ at which there is no longer any expectation that the true point E, N may exist at that location. A TFC does not have to be symmetrical about the PME. Theoretically a TFC would define a rectangle (or square) over which $\mu_E(x)$ and $\mu_N(x)$ vary continuously. In practice, however, and in the nature of coordinate systems, only the axes of the two fuzzy numbers making up the TFC need be considered.

Two triangular fuzzy coordinates will subtend a fuzzy line (Figure 13.3(a)). Such a line has three components. The first is a *line of maximum expectation* (LME) subtended by the two points of maximum expectation in the TFC so that $\mu_L(x) = 1$. On either side of the LME will be two lines subtended by the extents of the TFC where $\mu_E(x) = 0$ or $\mu_N(x) = 0$, whichever gives the greatest perpendicular distance from the LME. These are *lines of zero expectation* (LZE) where $\mu_L(x) = 0$ and form the outer extents of the fuzzy line. The LZE and LME need not be parallel nor need the LZE be symmetrical on either side of the LME. It can be assumed for simplicity that expectation declines linearly from the LME to the LZE and that together they form a *triangular fuzzy line*. In other words, for both the point and the line, the key known features are where membership values $\mu_A(x) = 1$ and $\mu_A(x) = 0$; once established, all else is linear interpolation. Rather than constructing some complex surface for the triangular fuzzy line, the assumption of linear interpolation is in response to the desire not to fall into the trap of being too precise about measures of uncertainty, and the natural wish for conceptual simplicity.

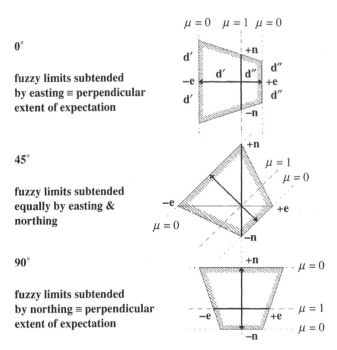

Figure 13.4   The isosceles trapezoid of triangular fuzzy coordinate extent.

Where a boundary is defined *per se* rather than from a consideration of discrete points, an alternative approach can be taken in defining a triangular fuzzy line (Figure 13.3(b)). The extents of $\mu_A(x) = 0$ for each section of boundary can be defined by a *perpendicular extent of expectation* (PEE) both to left and right of the LME. Thus the PEE signifies the distance on either side of the LME over which there is an element of belief that the true line may exist, but diminishing linearly to zero at the furthest extent of the PEE. The PEE may be asymmetric about the LME, reflecting differential levels of purity in left and right polygons. The possibility of asymmetry gives greater flexibility in boundary representation than, say, global measures of ε.

For the purposes of computational geometry, there is a need to have a technique whereby triangular fuzzy coordinates of points can be derived from a triangular fuzzy line. There are *three* cases that need to be considered here:

1   Where a triangular fuzzy line is subtended by two sets of triangular fuzzy coordinates, the fuzzy extents of any point along that line are based on the linear interpolation of the fuzzy extents of the two TFC subtending the line. Such a point is illustrated in Figure 13.3(a).

2   Where the fuzzy extents of a triangular fuzzy line are determined by perpendicular extents of expectation, it is not possible to simply extend the fuzzy extents of any new point to intersect the lines of zero expectation. The problem arises that, when the direction of the line is along one the coordinate axes, the fuzzy extents of either the easing TFC or the northing TFC are parallel to the LZE and would run on for infinity. To overcome this problem, an isosceles trapezoid rotated in the direction of the line is assumed to surround the new point (Figure 13.4). The dimensions are determined by the PEE. The fuzzy extents of the TFC are calculated by the intersection with the isosceles trapezoid. Maximum extent of the TFC is at 45° when it is PEE × √2. This

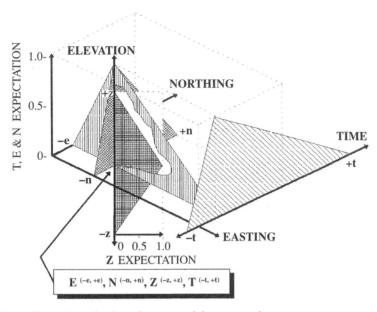

Figure 13.5 Illustration of a four-dimensional fuzzy coordinate.

operation and that of (1) above can be carried out with basic computational geometry used in computer graphics and vector GIS (e.g. Preparata and Shamos, 1985).

3 Where two or more triangular fuzzy lines intersect as in overlay, the TFC for the point of intersection is first calculated separately for each of the lines. These are then combined using a MAX function (union) in which the furthest extents in each of the cardinal directions is taken. This is likely to lead to locally-increased fuzzy extent for all lines forming the intersection, but no different from what would occur, say, when combining the root mean square error (RMSE) of the points.

Having established the concepts of triangular fuzzy coordinates and subtended lines it is possible to move up one more dimension to *fuzzy polygons* defined by a series of connecting triangular fuzzy lines having closure. Data-manipulation can follow standard computational geometry for any point and line of maximum expectation and for lines of zero expectation according to rules given above. Thus the coordinate system can be transformed or used in overlay analysis in very much the traditional way.

Since a triangular fuzzy extent, whether in a point or in a line, gives the full extent of expectation from $\mu_A(x) = 1$ to $\mu_A(x) = 0$, it can be taken as equivalent to three standard deviations of the normal distribution. This allows accuracy measures quoted as $\pm\sigma$ to be represented by equivalent fuzzy numbers. This, in turn, permits the integration of certain global accuracy measures alongside specifically-indicated boundary sharpness which represent transition zones or other forms of uncertainty.

A fuzzy coordinate can itself be extended to higher dimensions along the same principles to include elevation and time which are also expressed as TFN (Figure 13.5). It should thus be possible to express approximate objects in a GIS that are time-dependent.

13.5 DATA STRUCTURES FOR HANDLING FUZZY COORDINATES

The form of data structure required for handling triangular fuzzy coordinates and lines only needs be an extension of what is currently used in vector GIS. For example, Figure

13.5 shows an entity-relationship (Chen, 1976) diagram for a typical two-dimensional topological data structure for a layer comprising of nodes, lines, segments and polygons. Node, line and polygon all have feature identification numbers (ID). Polygons would normally have a user-defined label or tag; tagging would also be available for lines and nodes, but not every node or line would be expected to be tagged. Coordinates are attributes of nodes and points making up the segments. Finally, left and right polygon ID are stored as attributes of a line. Figure 13.5 shows this data structure further modified for triangular fuzzy coordinates and triangular fuzzy lines. The change is minimal with the addition of fuzzy extents $E^{(-e, +e)}$, $N^{(-n, +n)}$ to the coordinates or, if used, left and right perpendicular extents of expectation for lines. One wouldn't normally expect to store both TFC and PEE, as one would subsume the other. For most applications only the PEE would be present, since most lines are likely to be made up of many segments, and it would be easier to handle fuzzy extent as an asymmetric buffer zone from the PEE rather than trying to interpolate linearly from the nodes to all the points constituting a line.

The effect of this additional information on disk and memory requirements needs to considered, although the cost of hardware has significantly decreased in recent years. The new information can be thought of as doubling the number of attributes at each node or increasing the line attributes by 50%. Whether or not this is significant, and thereby offsets perceived advantages to be had from the presence of quality data, is difficult to assess. It depends largely on the proportion of lines and nodes in the map layer and particularly the proportion of data forming segments of lines. But it is clear that recording PEE rather than TFC is likely to be more compact. If left and right PEE are recorded as 8 bits each (0–256 metres), then an experimental layer with 4 689 lines forming 1 688 polygons (fully topological) would require 9.4 K bytes or 1% increase on its current file size. Another layer of 693 lines forming 339 buildings would require an additional 0.7% storage. There could conceivably be no storage overhead if bit packing using struct in C language was employed. For example, the coordinates used in the experimental data layers could be stored in 20 bits, leaving 12 bits spare in a 32-bit architecture for the fuzzy extents, i.e. up to $E^{(-64, +64)}$ with sign implicit. C code would be as follows:

```
struct fuzzy_e {
        unsigned int easting:20
        unsigned int minus_e:6
        unsigned int plus_e:6
}
struct fuzzy_n {
        unsigned int northing:20
        unsigned int minus_n:6
        unsigned int plus_n:6
}
```

Thus a triangular fuzzy coordinate $E^{(-e, +e)}$, $N^{(-n, +n)}$ could be stored in two 32-bit numbers, in the same way as traditional coordinates E, N. Uses of struct for packing the quality data into the spare parts of bytes can also be extended to the left and right PEE of triangular fuzzy lines. These are shown in Figure 13.6. Thus, depending on implementation strategy, the quality data could be stored with no disk overhead. There is necessarily a loss of performance speed, however, when using struct, and this would need to be borne in mind when considering the trade-off between storage and performance.

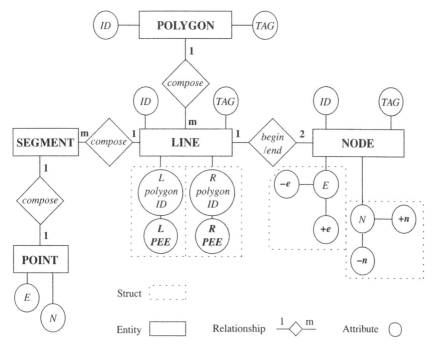

Figure 13.6 An entity relationship model for a two-dimensional topological data structure with the addition of triangular fuzzy coordinates and lines.

13.6 CONCLUSIONS

A GIS can only represent a boundary as being sharp if there is no data to support alternative interpretations. This paper has put forward the concept of fuzzy geometric entities based on triangular fuzzy coordinates as an alternative to crisp coordinate geometries as a means of recording boundary uncertainty explicitly. The coordinate system can be extended into four dimensions to include elevation and time. A coordinate geo-metry can be developed that satisfies the needs of GIS data-handling. The data structure can also be compact, so that few overheads may accrue from the inclusion of the add-itional data. This offers opportunities in GIS applications, where boundaries are inherently inexact, to have more representative data that better reflect the true situation.

References

ALTMAN, D. (1994) Fuzzy set theoretic approaches for handling imprecision in spatial analysis, *Int. J. Geographical Information Systems*, **8**, 271–289.

BLAKEMORE, M. (1984) Generalization and error in spatial data bases, *Cartographica*, **21**, 131–139.

BOLSTAD, P., GESSLER, P. and LILLESAND, T. (1990) Positional uncertainty in manually digitized map data, *Int. J. Geographical Information Systems*, **4**, 399–412.

BRIMICOMBE, A.J. (1997) A universal translator of linguistic hedges for the handling of uncertainty and fitness-for-use in Geographical Information Systems, in Kemp (Ed.), *Innovation in GIS 4*, 115–126, London: Taylor & Francis.

BURROUGH, P.A. (1986) *Principles of Geographical Information Systems for Land Resources Assessment*. Oxford: Clarendon Press.

BURROUGH, P.A., MACMILLAN, R. and VAN DEURSEN, W. (1992) Fuzzy classification methods for determining land suitability from soil profile observations and topography *J. Soil Science*, **43**, 193–210.

BURROUGH, P.A., VAN RIJN, R. and RIKKEN, M. (1996) Spatial data quality and error analysis issues: GIS functions and environmental modelling, in Goodchild *et al.* (Eds), *GIS and Environmental Modelling: Progress and Research Issues*, 29–34, Fort Collins CO: GIS World Books.

CHEN, P.P. (1976) An entity-relationship model: toward a unified view of data, *ACM Transactions on Database Systems*, **1** (1), 9–35.

CHRISMAN, N.R. (1982) A theory of cartographic error and its measurement in digital databases, *Proceedings AutoCarto*, **5**, 159–168.

CHRISMAN, N.R. (1983) Epsilon filtering: a technique for automated scale changing, *Technical Papers, 43rd Ann. Meeting of the American Congress on Surveying & Mapping*, 322–331, Washington DC.

DRUMMOND, J.E. (1987) A framework for handling error in geographic data manipulation, *ITC Journal* 1987–1, 73–82.

DUNN, R., HARRISON, A.R. and WHITE, J.C. (1990) Positional accuracy and measurement error in digital databases of land use, *Int. J. Geographical Information Systems*, **4**, 385–398.

EDWARDS, G. and LOWELL, K.E. (1996) Modelling uncertainty in photointerpreted boundaries, *Photogrammetric Engineering & Remote Sensing*, **62**, 377–391.

GOPAL, S. and WOODCOCK, C. (1994) Theory and methods for accuracy assessment of thematic maps using fuzzy sets, *Photogrammetric Engineering and Remote Sensing*, **60**, 181–188.

KOLLIAS, V. and VOLIOTIS, A. (1991) Fuzzy reasoning in the development of geographical information systems, *Int. J. Geographical Information Systems*, **5**, 209–223.

LAW, J.S.Y. and BRIMICOMBE, A.J. (1994) Integrating primary and secondary data sources in Land Information Systems for recording change, *Proc. XX FIG Congress*, Melbourne, Vol. 3, 478–489.

MAFFINI, G., ARNO, M. and BITTERLICH, W. (1989) Observations and comments on the generation and treatment of errors in digital GIS data, in M.F. Goodchild and S. Gopal (Eds), *Accuracy of Spatial Databases*, 55–67, London: Taylor & Francis.

PREPARATA, F.P. and SHAMOS, M.I. (1985) *Computational Geometry*, New York: Springer-Verlag.

ROBINSON, V.B. and FRANK, A.U. (1985) About different kinds of uncertainly in collections of spatial data, *Proc. AutoCarto 7*, 440–449, Washington D.C.

UNWIN, D. (1995) GIS and the problem of 'error and uncertainty', *Progress in Human Geography*, **19**, 549–558.

WANG, F. and HALL, G.B. (1996) Fuzzy representation of geographical boundaries in GIS, *Int. J. of Geographical Information Systems*, **10**, 573–590.

ZADEH, L. (1965) Fuzzy sets, *Information and Control*, **8**, 338–353.

ZHANG, G. and TULIP, J. (1990) An algorithm for the avoidance of sliver polygons in spatial overlay, *4th Int. Symp. on Spatial Data Handling*, Zurich, Vol. 1, 141–150.

CHAPTER FOURTEEN

Supporting complex spatiotemporal analysis in GIS

HOWARD LEE AND ZARINE KEMP

14.1 INTRODUCTION

Geographical information systems (GIS) are currently presented with notable challenges by applications in environmental modelling and management. These applications are characterized by large heterogeneous spatiotemporal datasets involving many variables. The spatial data infrastructure is often a subject-oriented collection of framework data as well as field data (Goodchild, 1996) used for a variety of monitoring and decision-support purposes. In these circumstances, supportive computing frameworks are needed to underpin spatiotemporal research in an exploratory and intuitive manner (Ahlberg, 1996; Densham, 1994; Mesrobian, 1996). Among other requirements, the computing framework should allow:

- The user to deal with natural representations of the phenomena and relationships of interest.
- The application of statistical functions to the raw dataset so that flexible partitioning of the problem space is achieved (Rafanelli, 1996).
- The intuitive and interactive visualization of selected subsets to enhance exploration and analysis.

Although existing GIS are able to capture changes by successive snapshots, they lack the ability to support complex analysis involving space-time as well as other dimensions (attributes) referred to in the literature as aspatial properties of a spatiotemporal object. By complex analysis, we refer to a non-trivial combination of data extraction from some data repository, and manipulation of that data. This paper describes a framework that is being developed in order to support analysis in domains with complex space-time characteristics.

14.2 AN ILLUSTRATIVE SCENARIO

This research is part of an initiative to develop a system for integrated fisheries and marine environmental management. It should be stressed that these concepts are by no means confined to this area. Successful management of the fishing industry transcends a purely quantitative approach towards stock assessment, since there are countless ways of achieving the same target stock level (Hilborn, 1992). Consider these two policies: (i) intensive fishing efforts followed by a recuperation period, or (ii) less intensive but year-round fishing. Although both approaches may produce similar target stock levels, there will be significant differences in their short- and long-term prospects for economic gain and fish population dynamics. Any system operating under such circumstances must provide flexible analytical support. The fundamental themes of space and time provide a common thread that runs through the various perspectives that relate to any given spatiotemporal system. These perspectives can best be observed as short-, medium- and long-term goals of the application (Meaden, 1997).

From the *short-term* perspective, the application provides functionality for the capture, retrieval and analysis of all data on fishing activities. Input data consist of species fished, quantities thereof, spatial location of fishing (related to specific management areas), dates or times when the activities took place, and the ownership or registration number of vessels that are participating in the fishery. This information is comprehensive enough to encompass a range of queries that arise with respect to marine fishing such as: *what? where? when? how much?* and *by whom?* Monitoring information here accommodates short-term interests.

In the *medium-term*, the application is used to retrieve summarized or aggregated information in a variety of ways. The application also provides for the inclusion of data on fishing quotas, so enabling management bodies to maintain control of this important regulatory aspect. The flexible model provides resource managers (such as the Ministry of Agriculture, Fisheries and Food in the UK) with sufficient information to support a range of policy- and decision-making processes. The following give an indication of typical queries in the medium-term:

- Total weight of species s_1 by boat owner b_1.
- Total weight of species s_1 by all UK registered boats.
- Summary of the weight of all species fished between time interval t_1 and t_2.
- Change in the total catch weight of species s_1 between time t_1 and t_2.
- Total catch per unit effort in user specified marine areas.
- Total catch weight of species s_1 in year y_1 compared with European Union fishing quotas.

From the *long-term* perspective, the application is used to integrate data related to fishing activities with environmental data such as physical measurements drawn from CDT (conductivity, depth and temperature) sensors, as well as remotely-sensed marine data. The goals are to enable long-term modelling of sustainable fish stocks of various species, and monitoring of changes in the marine environment. Examples of these objectives include:

- Establishing the 'normal' range of a fish stock and the measurable fluctuations in this range.
- Measuring the degree to which catches are matched to a range of local physical environmental indicators.
- Establishing the relationship between fish concentrations and spatial and temporal variations in food supply and type.

- Measuring the disturbance to the sea bed ecology by trawling activities.

- Studying the spatial variability in fish community structures.

- Selecting locations for fish stock enhancement programmes using the knowledge gained from optimum fish biomass environments.

14.3 THE SPATIOTEMPORAL FRAMEWORK

14.3.1 Rationale for the system

In a multi-user computation modelling system, each user will build models associated with the task at hand. Analysts are likely to pose complex queries involving space, time and other dimensions; for example, 'How much has this country's fishing area grown over the last decade?' or 'Which direction has this school of fish been moving?' There are of course many other types of spatiotemporal processes (Claramunt, 1995; Lhotka, 1991; Peuquet, 1994). Our concern is to provide the analyst with a general capability for specifying the patterns and aggregates that may be required across several dimensions. This is an important issue that has been generally neglected by spatiotemporal and temporal database communities (Ozsoyoglu, 1995; Schmidt, 1995; Tansel, 1993).

Generally, different requirements engage different views from a particular spatiotemporal data set. Moreover, in a multidimensional problem space the constraints and conditions that apply to each dimension can only be expressed in the context of the analytical process being carried out. For illustrative purposes, let us consider the range of requirements for the temporal dimension.

Despite the seemingly arbitrary nature of all dimensions, two common properties can be identified:

- Multiple granularity.

- Multiple abstraction.

- When a map is explored at various scales, corresponding degrees of detail are revealed; similarly, temporal information can also be presented at differing levels of *granularity*[1] (e.g. daily, weekly or yearly) with each level of granularity exposing distinct patterns and trends. For example, examining the characteristics of a shoal of fish at hourly intervals, we may discover how these fish behave according to whether it is day or night, whereas if we extend our granularity to that of monthly intervals, we might better understand the migration patterns of the species. Hence, granularity is one of the parameters which determine the outcome of an analysis.

- Moreover, when data is examined at a coarser temporal granularity than it was recorded, the data needs to be *aggregated* accordingly. On the other hand, if the water temperature of some location was sampled at hourly intervals but viewed at a granularity of a day, the values will need to be averaged out. If we were interested in comparing the total weight of fish harvested by a trawler over several periods, the values would be summed instead.

- Another characteristic that emerges is that of temporal *inhomogeneity*. If, for example, the reproductive cycle of a particular species of fish is being considered, periods of heightened activity are of greater relevance to the analysis than periods of relative inactivity. Similar requirements for context-dependent temporal rules and user-defined semantics for the manipulation of temporal variables have also been noted in other applications (Chandra, 1995).

- When dealing with change, some spatiotemporal processes can be categorized as *event-based* (Peuquet, 1995) and others as *continuous*. Event-based processes include fishing vessels embarking on fishing trips and trawling for fish at particular locations, whereas changes in variables such as temperature, salinity and biomass concentration may be described as continuous. Data-collection techniques render continuous processes as discrete events and sampling intervals may not be regular. The issue here concerns the accurate reconstruction of various phenomena.

The complexity inherent in dealing with time also occurs in all the other dimensions involved in spatiotemporal analysis. Consequently, the underlying framework has to be able to deal with application-dependent irregularities in the data retrieved and analysis procedures.

14.3.2 The analytical abstraction layer (AAL)

In the light of the factors presented in the previous section, we propose the *analytical abstraction layer* (AAL) (see Figure 14.1) for characterizing how data is abstracted. AAL functionality includes aggregation, classification, generalization, characterization and partitioning to enable patterns and anomalies alike to be elicited. Given a query from an event-based model, AAL would interpolate necessary values to produce a dataset suited to some given application requirement.

AAL views data as a multidimensional entity. Time, location, catch weight and biomass densities are examples of what a dimension can represent. Figure 14.2 illustrates an example of a data *partition*[2] through a given region, time and fishing vessel, with the catch in tonnes as the dimension of interest. There can be several dimensions of interest; for example, the spatial dimension may be overlaid with circles of varying sizes denoting the amount of catch. Figure 14.3 shows the same data as Figure 14.2 but at a coarser granularity – the southwesterly and southeasterly regions are aggregated according to the seasons. One can also envisage a single aggregate value of all the regions and vessel

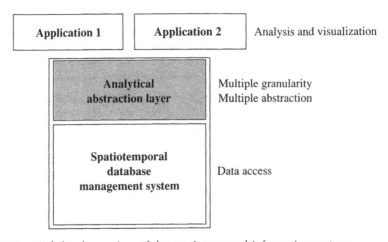

Figure 14.1 High level overview of the spatiotemporal information system.

North east
South west
South east
region
time
vessel type
dimension of interest: catch (tonnes)

	Jan	Feb	Mar	...	Dec
Class I	0.32	0.33	0.28	...	0.25
Class II	0.35	0.33	0.31	...	0.28
Class III	0.42	0.45	0.32	...	0.29

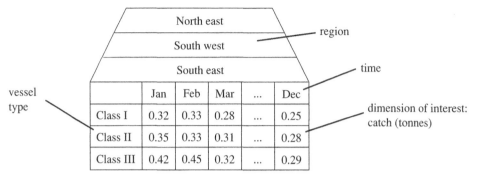

Figure 14.2 Catch values sorted according to month and vessel type in a given region.

North
South

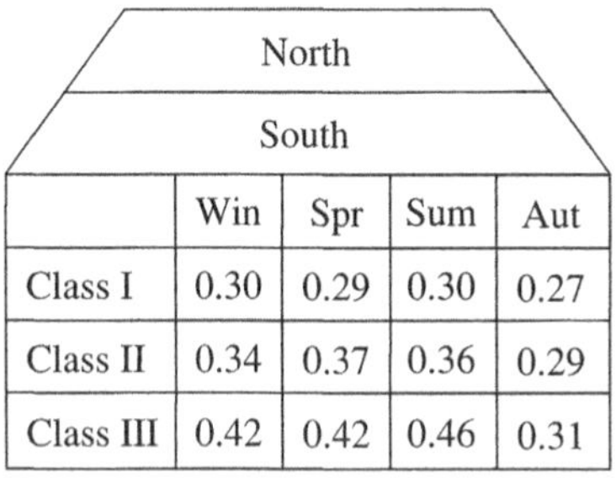

	Win	Spr	Sum	Aut
Class I	0.30	0.29	0.30	0.27
Class II	0.34	0.37	0.36	0.29
Class III	0.42	0.42	0.46	0.31

Figure 14.3 Catch averages over the seasons sorted according to vessel type.

0.30 > catch ≥ 0.45
catch > 0.45
Spr
Class III

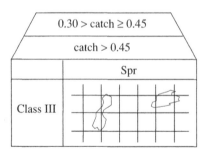

Figure 14.4 Regions denoting catches of greater than 0.45 tonnes (average) during spring.

types over the entire year. Figure 14.4 pivots the dimension of interest from catch weight to the spatial dimension.

Despite the seemingly arbitrary nature of all dimensions, two common properties can be identified:

- Multiple granularity.
- Multiple abstraction.

These two properties establish the foundations of AAL, and between them provide support for a wide number of analytical procedures. The following subsections provide an overview.

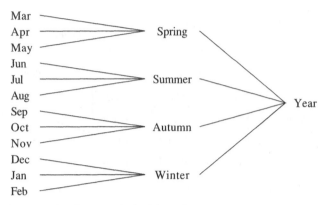

Figure 14.5 An example of a granularity hierarchy.

Multiple granularity

Changing the granularity of our data can change the interpretation of spatiotemporal phenomena. It also helps one generalize less important information by grouping them into collections that are more manageable. Each dimension defines a *granularity hierarchy*, where each node represents a different granularity of that dimension. Although this is an arbitrary procedure, many hierarchies are likely to be reused in different contexts such as the template depicted below (Figure 14.5).

When a dimension changes granularity, the values given by the dimension of interest changes according to its associated *granularity operator*. When the temporal granularity in Figure 14.2 of months changed to seasons in Figure 14.3, the dimension of interest, 'catch' is averaged out accordingly. In other circumstances, the dimension of interest could be summed instead. All granularity operators belong to one of the two categories (also see Figure 14.6):

■ *Aggregation*: data granularity becomes coarser.

■ *Decomposition*: data granularity becomes finer.

Figures 14.2, 14.3 and 14.4 illustrate an aggregation operator, or more specifically, the elementary averaging function. In other circumstances, summation and count operators may be more appropriate, for example, as if we were interested in the total catch over a year, or the number of months with a catch level of less than 0.36 tonnes. Some examples of a few predefined operations include:

■ *Aggregation operators*. Sum, average, count, difference, spatial generalization.

■ *Decomposition operators*. Linear interpolation, *b*-spline interpolation, spatial interpolation.

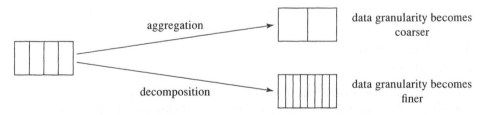

Figure 14.6 Granularity operators.

In some cases, where the dimension of interest is spatial, the change of granularity of other dimensions may result in larger spatial regions being retrieved. For display purposes, it may be sensible to rescale the regions. Spatial generalization techniques may be used to suppress unnecessary details. If spatial granularity was made finer than its recorded resolution, recent developments suggest it is possible to derive values by kriging or using techniques exploiting spatial dependency (Atkinson, 1996).

Multiple abstraction

Multiple abstraction is very useful for data processing and organizing data meaningfully as well as expressively. Data sets may belong to one or more *abstraction classes*. Each abstraction class in turn can have many members. Greenwich Meridian Time (GMT) and Pacific Standard Time (PST) are examples of abstraction classes. Data sets from different abstraction classes can only be compared subject to a *mapping function*. GMT and PST events are comparable only after the necessary adjustments are made (PST data is 8 hours behind GMT). Mapping functions may of course be arbitrarily complex, and are especially useful in reconciling incomparable data sets caused by uncalibrated instruments or from socio-economic factors like inflation.

For simplicity, one can impose the restriction requiring all members of an abstraction class to have the same granularity. It is only permissible to allow different granularity if mandatory mappings between the different granularity levels are defined. This is important if one wants to be able to apply a common set of query operators, such as those involving topology, on the data sets. This issue is akin to correlating a landmark on two maps of the same area but at different scales.

Multiple abstraction also allows the same data set to be portrayed in different forms. For instance, spatial data can be represented in a raster or vector format. Temporal data can be represented in time or frequency, where frequency is the reciprocal of time (Figure 14.7). Frequency domain analysis is especially effective for eliciting cyclic phenomena. Different representations emphasize different aspects of the data. Each abstraction class also defines *access methods* for data extraction as well as various functions and metadata belonging to the abstraction class. Access methods also encapsulate granularity hierarchies and the relevant operators. An access method may take input parameters like the level of granularity and range of data. Once a mandatory number of access methods are invoked, the relevant data is then retrieved by the abstraction class and appropriately displayed by a separate visualization module.

Figure 14.8 shows the main elements of an abstraction class definition for a fish species informally known as capelin. This is essentially an *object-oriented* approach to the management of data for the purposes of analysis and, in particular, exploratory analysis. This approach provides a modular management scheme for data sets as well as enabling the re-use of common definitions.

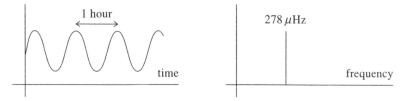

Figure 14.7 Time and frequency (amplitude only) representations of the same data set.

```
abstraction class = fish species
     dimensions = location, time, quantity

  access methods (including granularity hierarchy and operators) =
     location (input parameters = granularity level, range etc.)
     time
     quantity
     frequency
     circular time
```

Figure 14.8 Sample overview of an abstraction class definition.

Certain abstraction methods *reduce* the number of dimensions or the complexity of a data set. Essentially, the mapping function under such conditions maps a dimension on to a narrower range of possible values or to some other existing dimension. One approach is *circular data* (Fisher, 1995). In many instances, circular data comes across much clearer (visually and analytically) than the more conventional linear form (Figure 14.9).

Spatial data involving directions like wind and ocean currents can also benefit from this approach. It is important to realize that circular data are not just visually different from linear data – observe the lack of start and end discontinuities, which is a particularly useful property in cluster analysis.

To illustrate AAL's principal features in a practical situation, let us examine an example. The scenario: there are a set of trawl paths, i.e. where fishing boats lower their nets into the water and, associated with each path, information about how much fish was caught, the types of fish, etc. Strictly speaking, the path actually consists of a set of sampled points from a global positioning device (GPS) and the time each was recorded. Assuming the samples are frequent enough, one can deduce the path taken by a vessel. The problem (see Figure 14.10): what if the analyst selects an area encompassing several partial paths to ask how much fish was caught in a particular region?

A simple approach may be to return the total catch of all the paths within the selected area. However, it is possible to be a little more accurate if we could assume the fish catch was evenly distributed over the trawled path. A more sophisticated approach, therefore, would calculate the amount of fish caught by a path to be the ratio of the path length

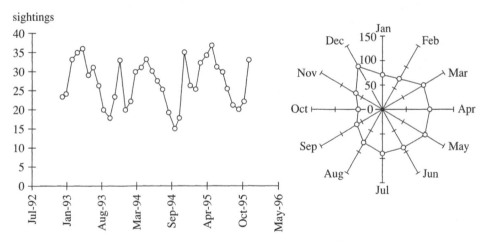

Figure 14.9 A linear plot and circular plot of a particular kind of fish sighted in a given region.

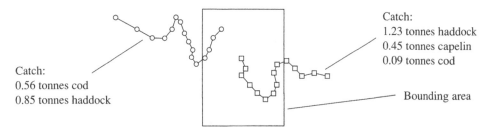

Catch:
1.23 tonnes haddock
0.45 tonnes capelin
0.09 tonnes cod

Catch:
0.56 tonnes cod
0.85 tonnes haddock

Bounding area

Figure 14.10   The total catch problem.

```
abstraction class = trawl path {
        dimensions = location, time, fish type, fish quantity

        access methods =
            simple_catch
                length_catch(bounding area)
            area_catch(bounding area)
        ⋮
        etc.
}
```

Figure 14.11   Abstraction class for the catch problem.

encompassed by the selected area to the total path length of that trawl. Another possibility is to define a bounding region of the trawl and calculate the amount of fish caught as proportional to the size of the overlapping area between the selected area and the bounding region. Other approaches could take into account the behaviour of the marine species and type of gear or net used.

This scenario can be managed by defining an abstraction class with the appropriate access methods, ignoring other details such as granularity (Figure 14.11). Note the three aforementioned variants of catch.

14.4   RELATED ISSUES

The research described in this paper is part of the emerging discipline sometimes re-fer-red to as *environmental informatics* which seeks to combine the areas of GIS, temporal database systems, modelling, simulation, visualization and knowledge-discovery. The work should be viewed in the context of the complete framework for modelling and knowledge-elicitation. By their nature, these activities are iterative and evolve by stages. To complete the picture, we briefly outline the other related modules of the framework.

Visualization is closely linked to multidimensional partitioning. The prototype system provides a flexible presentation mechanism that enables users to determine the combination of display and visualization required – e.g. cartographic, graphical and tabular. Later versions of the system will make the visualization module as generic as the data-partitioning and retrieval processes, with implicit support for representing change and built-in mechanisms dependent on the primitive object types involved. The importance of visualization as a means to explore and understand geospatial phenomena has been considered by other researchers too (Ahlberg, 1996; Egenhofer, 1994; Keim, 1996).

Another related module concerns the capture and persistent storage of any derived multidimensional partitions. The problem is obviously of importance in environmental informatics where datasets are very large and possibly distributed, and analysis progresses in stages. The issues to be tackled related to the modelling of the multidimensional structures – their physical representation and capabilities provided for manipulation. Being multidimensional, they do not easily map into the models underlying standard relational database systems because of their different structural and behavioural characteristics. One solution is to use the extensible typing features of object-oriented or object-relational systems. The framework of this research adopts an object-relational platform, Illustra (Informix Software®), of which Weldon (1997) provides an overview. This permits *both* field data and derived multidimensional partitions to reside in an integrated object repository.

14.5 CONCLUSIONS

We have endeavoured to show the intricacies of analysis in the spatiotemporal domain by an account of the generic retrieval requirements and propose a framework. An important element of the framework is the analytical abstraction layer or AAL, which provides better management and understanding of spatiotemporal problems. Data in AAL is multidimensional and all dimensions are characterized by multiple granularity and multiple abstractions. This means we can view data at various levels of detail and in a multitude of representations. In turn, the information we retrieve can be used for complex analysis and for constructing higher order patterns of spatiotemporal processes.

With the focus on spatiotemporal information systems as opposed to spatiotemporal databases, we have taken a step towards realizing the enormous potential of such systems. The framework is particularly effective at exploratory analysis whereby different types of data are used interactively and collaboratively by analysts. Although analytical techniques have traditionally been associated with statistics, the rapidly growing load of spatiotemporal data from sources like earth observation systems are calling for an approach to analysis resembling our process of *knowledge discovery*. However, in order to support knowledge discovery, it is important for different types of data to be able to exchange information in a standardized manner. Knowledge discovery is essentially a search for *interesting* patterns. The user-mediated mechanisms for organizing the search process ensures that any discovered knowledge during analysis is consistent with the user's requirements. This also helps constrain the search process with domain knowledge not easily available within the system. We believe this framework will play an important role in helping to develop the next generation of geographical information systems.

Acknowledgements

Howard Lee is funded by an E.B. Spratt bursary awarded by the Computing Laboratory, University of Kent at Canterbury.

Notes

1 The term 'granularity' is akin to the notion of resolution, which is arguably more precise. In order to attain a consensus over the confusing terminology found in early temporal database research, 'granularity' has been selected in preference over other terms. Nevertheless, its application context should be clear enough to avoid any ambiguity.

2   In this context, the term 'partition' refers to a multidimensional and possibly non-contiguous subset of the data.

References

AHLBERG, C. (1996) Spotfire: An Information Exploration Environment, *SIGMOD Record*, **25** (4), 25–29.

ATKINSON, P.M. (1997) Mapping sub-pixel vector boundaries from remotely sensed images, *Innovations in GIS 4*, London: Taylor & Francis.

CHANDRA, R., SEGEV, A. and STONEBREAKER, M. (1995) *Implementing calendars and temporal rules in next generation databases*, Technical report from Lawrence Berkeley Laboratory, University of California. http://epoch.cs.berkeley.edu:8000/postgress95/.

CLARAMUNT, C. and THERIAULT, M. (1995) Managing Time in GIS: An Event-Oriented Approach, Int. Workshop on Temporal Databases, in A. Tuzhilin and J. Clifford (Eds), *Recent Advances on Temporal Databases*, Berlin/London: Springer-Verlag (published in collaboration with the British Computing Society).

DENSHAM, P.J. and ARMSTRONG, M.P. (1994) A heterogeneous processing approach to spatial decision support systems, in T. Waugh and R. Healey (Eds), *Advances in GIS research*, London: Taylor & Francis.

EGENHOFER, M.J. (1994) Spatial SQL: A query and presentation language, *IEEE Transactions on Knowledge and Data Engineering*, **6** (1), 86–95.

FISHER, N.I. (1995) *Statistical analysis of circular data*, Cambridge: Cambridge University Press.

GOODCHILD, M.F. *et al.* (Eds) (1996) *GIS and Environmental Modelling: Progress and Research Issues*, Fort Collins CO: GIS World Books.

HILBORN, R. and WALTERS, C. (1992) *Quantitative fisheries stock assessment: Choice, dynamics and uncertainty*, New York: Chapman and Hall.

KEIM, D.A. (1996) Pixel-oriented database visualisations, *SIGMOD Record*, **25** (4), 35–39.

LHOTKA, L. (1991) Object-oriented methodology in the field of aquatic ecosystem modelling, in J. Bezivin and B. Meyer (Eds), *Technology of object-oriented languages and systems (TOOLS)*, **4**, London/New York: Prentice Hall.

MEADEN, G. and KEMP, Z. (1996) Monitoring fisheries effort and catch using a geographical information system, *Developing and sustaining world fisheries resources: The state of science and management*, Collingwood: CSIRO Publishing.

MESROBIAN, E. *et al.* (1996) Mining geophysical data for knowledge, *IEEE Expert*, **11** (5), 34–44.

OZSOYOGLU and SNODGRASS (1995) Temporal and real-time databases: A survey, *IEEE Transactions on Knowledge and Data Engineering*, **7** (4), 513–532.

PEUQUET, D. (1994) It's about time: A conceptual framework for the representation of temporal dynamics in geographical information systems, *Annals of the Association of American Geo-graphers*, **84** (3), 441–461.

PEUQUET, D. and DUAN, N. (1995) An event-based spatiotemporal data model (ESTDM) for temporal analysis of geographical data, *Int. J. Geographical Information Systems*, **9** (1), 7–24.

RAFANELLI, M. *et al.* (1996) The aggregate data problem: a system for their definition and management, *SIGMOD Record*, **25** (4).

SCHMIDT, D. and MARTI, R. (1995) Time Series, a Neglected Issue in Temporal Database Research? International Workshop on Temporal Databases, in A. Tuzhilin and J. Clifford (Eds), *Recent Advances on Temporal Databases*, Berlin/London: Springer-Verlag (published in collaboration with the British Computing Society).

TANSEL, A.U. *et al.* (1993) *Temporal databases: Theory, design and implementation*, Redwood City: Benjamin-Cummings.

WELDON, J.L. (1997) RDMSes get a make-over: Are technologies such as datablades solutions or bandages for complex data management? *Byte Magazine*, **22** (4), 109–114.

CHAPTER FIFTEEN

A control strategy for automated land use change detection: an integration of vector-based GIS, remote sensing and pattern recognition

GARY PRIESTNALL AND ROGER GLOVER

15.1 INTRODUCTION

Remotely-sensed imagery has for many years provided a means of classifying land cover over wide areas and at regular temporal intervals, and has, therefore, enabled the detection of broad changes in patterns through time, largely at the pixel level (Cope Bowley and Piper, 1995). Such imagery also has the potential for detecting more detailed instances of change through identifying objects in a scene such as buildings and roads, thereby creating the possibility of updating cartographic databases. The problem is that, until very recently, commercially-available satellite imagery was limited to a maximum spatial resolution of 10 m, which led to difficulties in extracting man-made features. This requirement for finer pixel resolutions when extracting components of land use was made clear by Blamire and Barnsley (1996). The imminent arrival of high resolution satellite imagery will allow more sophisticated, automated, object-extraction techniques to be implemented, where objects with characteristic shape, dimensions and interrelationships become the focus of attention rather than pixels.

The focus of the work is, therefore, to detect instances of land-use change. Rather than work with imagery unaided, the intention is to explore the extent to which existing GIS vector databases can be used to guide change detection by confirming the presence of objects within the image. Central to this concept is the definition of scene objects such as buildings and roads. In order to detect change between the existing vector database and a new image, the vector and raster representations of these scene objects must be compared. Contextual knowledge given by the vector database, plus additional parameter information derived from this matching stage, may then be used to aid recognition of 'new' objects within the image.

To summarize, the following research questions are addressed by this work:

- To what extent can *a priori* knowledge from GIS vector map data help us automatically understand the scene and, therefore, what is the potential for automated land-use-change detection using this technique?

- To what extent can knowledge gained from this integration help automate the recognition of new objects in the scene by fine-tuning parameters used by object-recognition models, and offer contextual information for new objects to limit the possible hypotheses that may be appropriate?

- How important is the level of structuring present in the GIS database, and will an object-oriented database offer the ideal mechanism for implementing a system such as this?

Given that the representation and recognition of scene objects is crucial to this work, it is first necessary to put the problem of geographical object-recognition into perspective.

15.2 THE NATURE OF THE PROBLEM

The successful recognition of geographic objects from imagery requires a more global understanding of the image than the local pixel-based approaches based upon spectral characteristics, as used in many land-cover classifications – an issue partially addressed in a review by Wilkinson (1996). Not only do the shape, dimensions and radiometric characteristics of an object become important, but also the overall context of that object within a scene becomes vital for resolving local ambiguities. In addition to pattern-recognition techniques for extracting objects from raster scenes, there is a need for a strategy to integrate low-level pattern information with the more global knowledge of the scene as a whole. The problem of deriving higher-level descriptions of objects from raster scenes is one which characterizes the field of machine vision. Over 30 years of research work in machine vision (summarized in Sonka *et al.*, 1993) offers many ideas which are directly relevant to the development of methodologies to extract geographic objects from imagery.

The complexity of the problems studied within the field of machine vision varies enormously (Figure 15.1). For some domains such as recognizing features in scanned engineering drawings (Priestnall, 1994; Priestnall *et al.*, 1996), many of the rules for pattern recognition can be defined clearly because the drawings are made to conform to specified standards. Geographic features represented on satellite images do not conform to such rigorous standards, and developing a strategy for recognizing the desired objects is more challenging. Many features appearing in an image, particularly elements of the natural landscape are ill-defined (Poulter, 1996) and their spatial relationships cannot be predicted or modelled easily. Certain environments, however, such as the urban fringes, contain objects with more symmetry, and with spatial relationships and other characteristics which can be modelled to some degree, albeit using a more limited and heuristic rule-base than many other domains in machine vision.

Many machine-vision applications involve a very controlled environment, where the objects under study are of a consistent size and shape, such as industrial object recognition (Zeller and Doemens, 1982). The difficulty with distinguishing geographic objects is that, without this *a priori* knowledge of 'what the computer is looking at', it is very difficult to define a rule-base for object recognition. Also, the objects in question are much more variable in size, shape and complexity than objects in other areas of machine vision.

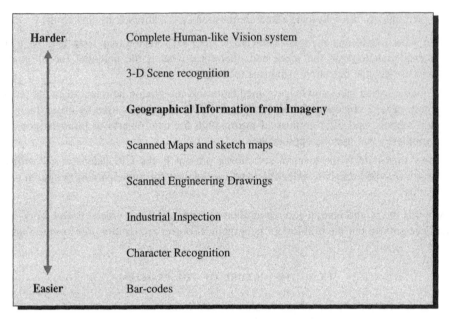

Figure 15.1 A generalized view of the relative complexity of some machine vision problems.

The automatic extraction of geographic objects from imagery for use within Geographical Information Systems can utilize various forms of sensors and, therefore, a variety of image types, monochrome, colour, multi-spectral, stereo pairs, and at a multitude of resolutions. The increasing requirement for vector data in more highly structured form within GIS will result in continued research interest in techniques for automating the identification of objects from images. Satellite platforms such as Landsat, providing pixel resolutions of around 30 m, offer very poor representations of many man-made objects such as roads and buildings, which are often of smaller dimensions than the pixels themselves. Research interest grew with the advent of the SPOT sensors, offering pixel resolutions of 10 m and allowing some success in feature-recognition, such as major roads (Guen *et al.*, 1995).

An alternative to satellite-borne sensors is aerial photography, increasingly available in digital form, and offering very high pixel resolutions, albeit suffering spatial distortions due to the relatively low altitude of the camera, and hence providing many research challenges when it comes to extracting geo-referenced objects. The extraction of building outlines from aerial imagery has received particular attention (McKeown *et al.*, 1985 and more recently Lin *et al.*, 1995), and increasingly the principles of photogrammetry are being exploited to derive complex 3-dimensional information from stereo imagery (McKeown *et al.*, 1996 and Kim and Muller, 1996). Much recent interest has been in the attempt to reconstruct 3-dimension urban models for planning purposes (Mueller and Olson, 1995), and particularly the use of high-resolution, digital elevation models derived from stereo pairs of aerial images (Weidner and Forstner, 1995). The challenges associated with the use of oblique imagery have been addressed by Cochran (1995).

Not all approaches are so complex, some being highly reliant upon spectral information alone (Shettigar *et al.*, 1995). In terms of identifying objects, the radiometric properties are considered important, but the size, shape and, vitally, the contextual information, both local and more global, are highly significant.

Recently there has been much interest in the imminent arrival of very high-resolution satellite data (Dowman, 1997). Potentially, this will offer good representation of shape as with digital aerial imagery, yet without many of the spatial distortions associated with photogrammetric data (such as buildings leaning away from the camera), and will also offer the wide area coverage and relatively frequent revisit times of satellite remote-sensing platforms. Missions are planned for 1997 onwards (Plumb, 1997) and, in the meantime, researchers can plan techniques to utilize this data using simulations.

15.3 APPROACH TO MAP-DRIVEN CHANGE DETECTION

The research work presented here aims to approach the problem of identifying objects in a slightly different way from that adopted by most conventional pattern-recognition methods. Rather than attempt to recognize objects in the image unaided, the intention is to integrate the imagery with the existing vector database for that area and to use the knowledge inherent in this vector model to assist the pattern-recognition process. Occasionally, the requirement for *a priori* knowledge about a scene is presented as a limitation of many machine-vision approaches, but there are many situations where digital maps exist in an area, and the potential for using these, both to detect change directly with the new image and to attempt to provide contextual clues which would otherwise have to be given by the user in a semi-automated system, should be investigated.

During the process of detecting change between objects in the GIS vector database and the image, the knowledge gained from this integration in terms of the typical parameters and locally consistent 'rules' can be used to check for the presence of new objects in the image. Many methods of object-construction from imagery such as the active contour models, or snakes used by many researchers (Gulch, 1995) require user interaction to give the necessary context for successful object extraction. Many semi-automatic, data-capture techniques require the user to enter seedpoints along a feature which can then be extracted, such as road lines (Gruen and Li, 1995), or line following software which is given parameters relating to the object type being extracted (such as Vtrak, Laser-Scan Ltd, 1996). The research question which arises from this part of the work is: to what extent can the vector objects be used to 'train' a system for extracting new objects from the image?

Map data have been used in other studies – for example, the use of existing vector polygon data to constrain the pixelwise classification of earth-observation data for land-cover monitoring, such as in the CLEVER mapping project (Smith *et al.*, 1997). Maitre *et al.* (1995) use raster road networks derived from scanned maps to suggest areas to search for road networks in an image, and Stilla (1995) and Stilla *et al.* (1995) use map data to limit the search space and increase the processing speed for aerial image understanding.

The work presented here is intended to use existing vector map data, not only to identify land-use change directly, but also to derive knowledge from this process, which can be used by subsequent object-recognition modules. It is felt that the success of this process will be reliant, to a large extent, upon the level of structuring and knowledge built into the vector database, and that a multi-level description of the map data would allow the best utilization of contextual information. To this end, an object-oriented GIS data model, as discussed later, is seen as providing the necessary knowledge content and flexibility to drive the image-analysis process.

The objects of interest in this study are those features typically represented on large-scale vector GIS data products such as the Ordnance Survey Landline data. The imagery currently being used is colour digital orthophotography at a spatial resolution of 0.5 m, degradable to resolutions of 1 to 5 m to simulate the likely resolutions of very-high-resolution satellite sensors. It must be stressed that this data is being used solely for algorithm development and to arrive at the framework of a system which will attempt to utilize very-high-resolution satellite data and appropriate-scale map data where available. Care has been taken to identify aspects of distortion in the digital photography which are likely to be less important in data from satellite sensors, such as building lean, and to choose test areas which are less affected by this. There is a large body of research aimed particularly at the use of photogrammetric techniques to extract both the horizontal and vertical edges of objects from imagery (McKeown *et al.*, 1996). Simulating satellite data from digital photography will not give a true representation of the likely character of various high resolution satellite products, and it is realized that certain stages of the process, such as the edge extraction described later, may need to be re-appraised.

The main study area is currently a small section of the University of Nottingham Campus, where known land-use change has occurred and can be surveyed on the ground. This has been the testbed for the development of techniques and algorithms, although it is intended that different environments will be tested when appropriate vector data and imagery becomes available.

15.4 A PROPOSED METHODOLOGY

A framework for the research is illustrated in Figure 15.2. The overall structure represents a balance between bottom-up information derived from the image and the vector data, and top-down control over the recognition of objects present in both data models. Two forms of change could be detected. First, vector features from the GIS database may no longer be present in the image and, secondly, new or modified objects, and therefore edges, may be present in the image, which are not present in the older vector dataset. In terms of bottom-up feature-extraction therefore, edge information must be extracted from the image and the appropriate vector features selected and structured.

The two forms of change necessitate two primary phases of the model (illustrated in Figure 15.2). The first is the confirmation of those objects which are present in the vector dataset, and the identification of weakly-confirmed objects which may in fact have been removed. The second phase is the identification of new objects which have no representation in the vector dataset. A high-level control strategy not only needs to resolve conflicting labellings of objects, but also to control the search for new objects using the clues or *a priori* knowledge supplied by the first phase.

The automated process of change-detection is initiated from the object database within the GIS, taking each object in turn, running a suite of macros to control the integration of the map vectors associated with each object with the edge information extracted from the image. The integration of the map vectors with the edges results in the vectors being attributed with information regarding the degree to which each object matches the image information – i.e. the likelihood that the map object is actually represented in the image. Also, the edges themselves are labelled during the process: any edges not corresponding to map objects are left unlabelled, and these will form the low-level primitives with which object-recognition modules will operate during Phase II of the process.

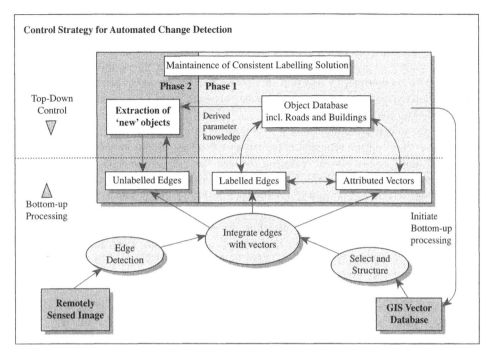

Figure 15.2 Outline of proposed methodology.

The exact order of the implementation of many elements of the study is largely dependent upon the nature of the available vector data. For example, building seed points or a more highly-structured vector database allows direct initiations of searches for buildings at certain locations. The information content and level of structuring of the vector data used to give *a priori* knowledge to identifying objects in the image is seen to be vitally important, and it is recognized that a highly-structured, object-oriented vector data model of the type described by Worboys (1994), would allow a more directly top-down approach to the first phase of the control strategy.

15.5 THE EXTRACTION OF EDGE INFORMATION FROM IMAGERY

The choice of methods to be used in the extraction of edges in this study was constrained to a degree by an early decision to use, where possible, only the functionality of the two chosen systems, namely: ESRI's ARC/INFO (7.0.3) and ERDAS'S IMAGINE (8.0). This decision meant that some of the commonly-used edge-detection filters used by other researchers such as the Canny–Petrou–Kittler filter (Kim and Muller, 1995) were not available for the current study.

These constraints led to the development of an edge-extraction methodology that treated building roofs as simple, straight-edged, planar surfaces that in an ideal image would appear as smooth homogeneous regions. The objective of the image-processing regime thus developed was to render these surfaces as smooth as possible, while maximizing the differences in mean pixel values between contiguous smoothed areas. This would create a series of steps in the image which could then be detected using a standard-edge detection filter.

The choice of smoothing filters was informed by the use of buffers either side of a number of manually-placed critical edges which had to be preserved during the smoothing process. These buffers, which provide descriptive statistics of the raster image to the 'left' and 'right' of each edge, allow the effect of different filters and stretches to be monitored by comparing pre- and post-processing figures. These statistics are also fundamental to later stages of the process, where they are used in the formulation and confirmation of building hypotheses.

It was found that an initial filtering with a series of median transects orientated along the x and y axes and the principal diagonals of the image produced reasonably smooth, noise-free regions. The use of transects reduced the risk of damaging thin lines and corners, problems associated with the use of square median windows (Sonka *et al.*, 1993, p. 75). A sigma filter (Abramson and Schwengerdt, 1993) with a threshold of 1.15 standard deviations was then passed over the image, producing a series of near-homogeneous regions. The application of a linear stretch to the processed image increased its contrast, helping the production of step edges in the image (Sonka *et al.*, 1993, p. 77). An initial set of edges was then produced using a Sobel filter (Mather, 1987, pp. 260–1). The sigma filter was passed over the image twice more with thresholds set at 3.3 and 4.45 standard deviations respectively, with edges being extracted after each pass.

The edge images were then exported from IMAGINE as ARC/INFO grids, where they were thresholded and reclassed into binary data sets. The resultant grids were then vectorized and combined using standard ARC/INFO commands to produce the final set of edges reproduced together with the raw image as Figure 15.3 (see colour section). Clearly the techniques just described work well with most of the building edges having been detected. Those areas where the process fails, A to E in the figure, are either very shallow edges where the contrast between the roof and its immediate surroundings is very low, or where an edge is still very noisy, or a combination of the two. The linear filters used in this study are incapable of addressing these problems. However, the application of a non-linear filter such as the lower-upper-middle (LUM) would help both reduce the noise and simultaneously steepen the edges in these problem areas. Hardie and Boncelet (1995) illustrate the efficiency of such filters when used in conjunction with a Sobel filter. Time constraints have meant that the potential of the LUM filter has not been investigated, but the authors believe that its incorporation into their system could yield significant improvements in the detection of these problem edges.

Prior to integration with the GIS vectors the edges had to be generalized to satisfy assumptions made by the subsequent matching process. The nearest function available in ARC/INFO implements the Douglas–Peuker line-simplification algorithm, which was shown to be unsuitable for use in this context. Given this fact, an algorithm based on a least squares solution that would generate two point lines that followed the detected edges more intuitively, whilst being sensitive to the preservation of corner details, was developed. At the time of going to press, this algorithm was being coded, and so the generalized lines shown here were produced manually.

Once generalized, the edges were processed automatically in ARC/INFO, using a module written in ARC/INFO's macro language (AML), where attributes such as an edge's orientation and statistics of the image pixels that lay to either side were attached to each edge. Similar attributes were recorded for the GIS vectors with each building and its constituent vectors given unique labels. The next section describes the procedure for attributing map vectors and edges, with information relating to the degree of correspondence between the map and the image.

15.6 INTEGRATION OF EDGES WITH GIS VECTORS

The aim here is to test the extent to which vectors in the GIS database are represented in the raster image, using any coincident, extracted-edge information together with other statistics derived from the image. The attributed vectors and edges were compared automatically in ARC/INFO using an AML module. The system records the relationships between each vector and the edges in its immediate vicinity – the search space being constrained by the use of buffers. Each vector is selected in turn, and the neighbouring edges are tested for near-parallelism, including the degree and type of overlap. This process is repeated until all the vectors have been processed.

Careful consideration was given to the possibility that the vector data might contain positional errors and that the imagery may display some geometric distortion, such as building lean. Regions of uncertainty, or tolerances, associated with the different possible relationships between each vector and its associated edges have been incorporated into the module.

Four main criteria currently contribute to the initial confirmation of either individual vectors or form important lower-level clues as to the confirmation of whole building objects. The criteria are as follows:

1 *Near-Parallelism.* If a vector and its associated edge(s) are near-parallel, their relationship is recorded – i.e. whether they are disjoint, nested, overlapping or equal (see Worboys, 1995, p. 182). The perpendicular distance from the mid-point of each parallel edge to its associated vector is also recorded, as is the percentage length of vector accounted for by near-parallel edges.

2 *Crossing Edges.* Edges that cross a vector are considered as strong evidence that change may have occurred and are used in the absence of near-parallel edges to reduce the likelihood of confirmation.

3 *Corners.* The process of extracting edges and their vectorization results in a loss of detail at building corners; rather than being right angles they tend to be mitres. This, however, is a consistent feature of this image-processing regime and, as such, can be used as further evidence of a building's presence within the image.

4 *Normal Lines.* In relation to many roof types, lines internal to the roof area, and normal to one or more roof edges, can provide additional supportive evidence that a building may still exist, provided the lighting and viewing geometries are favourable to the detection of such edges. These edges can be important characteristics of certain buildings in imagery, but are not usually represented by vector map data.

Once recorded, these relationships are used to produce an index of coarse probabilities indicating whether or not change may have occurred at the level of individual vectors. These, in turn, can be aggregated at the building level and used to categorize either individual vectors or buildings on a 5-point scale. An initial crude labelling scheme for individual vectors is given below in Table 15.1.

Corners and normal edges are considered factors that are important clues for the confirmation of a building as a whole, and their presence in appropriate places adds weight to the general confirmation given for a building by its individual map vectors. These initial labellings can then be used to provide contextual information that will aid the search for buildings present in the imagery but absent from the vector database.

Table 15.1 Initial labelling scheme for confirmation of map vectors.

% length of map vector accounted for		Crossing lines	Likelihood of change
if line < 10 m	if line > 10 m		
40–100	70–100	No	Very Likely (5)
25–40	30–70	No	Likely (4)
10–25	10–30	No	Possible (3)
0–10	0–10	No	Unlikely (2)
0–10	0–10	Yes	Very Unlikely (1)

Figure 15.4 (see colour section) illustrates the map vectors of a small sample of buildings manually selected from the test area and representing a range of conditions. The labels correspond to the likelihood of change which has been allocated by the automated procedures as described above. Several of the buildings are allocated a 'very likely' status, showing a direct correspondence between map vectors and image edges of over 80% of the total perimeter length involved.

15.7 THE INTRODUCTION OF CONTEXT AND A CONTROL STRATEGY

So far each map object has been compared to the imagery in isolation in an attempt to confirm its presence in the imagery. This has made use of the immediate context given by the map vector objects. At this stage we have a set of partially-confirmed map vectors and a set of partially-labelled image edges. The partially-labelled set of image edges will contain much ambiguity and will include many unlabelled edges, such as shadows of buildings, rather than real-world objects. What is required is a system to control the construction of a higher-level understanding of the objects present in the scene, each object being composed of several lower-level edges. The problem is one of allocating edges to higher-level objects, which may in turn affect the potential labels which are considered possible for neighbouring edges. As more potential objects are recognized, labelling conflicts may occur – i.e. two lower-level edges may be part of two or more possible labelling solutions. A large body of relevant work exists in the machine-vision literature, specifically work on recognizing sketch maps using a technique known as arc-consistency (Mackworth *et al.*, 1985) to repeatedly review the local labels given in an area, using rules defining the 'legal' combinations of local labellings, in order to arrive at a globally consistent solution. Also of relevance is the concept of a semantic network, as discussed in Sonka *et al.* (1993), which represents relative spatial interactions between components of a scene rather than relying on any kind of rigorous connected geometry.

The approach to be adopted in the current research involves attempting to 'fill' object classes of various types with bottom-up information derived from the image. The context given by the map data suggests the most likely object-hypotheses to try first. A cyclic process of testing 'legal' juxtapositions of recognized objects will attempt to maintain a more globally-consistent solution (Priestnall, 1994), although the rule-base used for this is less complete and well defined than other domains of machine vision. It is possible that conflicts may occur between the labelling of higher-level features such as buildings – for example, one edge may become part of several competing labelling solutions, in which case the data structure storing the 'conflict list' attempts to resolve the situation by choosing the most likely object description from the competing evidence. Research is ongoing in assessing the most appropriate data structure to store information about the current state of image labelling and where conflicts are occurring. Many vision systems employ a 'blackboard' (Stilla, 1995), a data structure to which conflicts are posted, and from which conflict-resolving functions are initiated.

Information from the first phase can, potentially, aid the extraction of new objects both through having developed a multi-level description of the map data and by having parameterized regional orientations, sizes and types of objects confirmed. Also, the crude nature and direction of shadows present in the imagery could be utilized. Finally, depending upon the exact object-extraction methods chosen, specific aspects of the parameterization of object-recognition models could be made. Attention is being given to the potential role of Artificial Neural Networks in the classification of buildings (Evans, 1996) within the overall strategy proposed here. Of particular interest is the extent to which the existing rule-base could form a suitable set of inputs to a neural network.

Currently the labelling of buildings has been a priority, but the ultimate aim is to use an object-oriented approach to data-modelling to include other objects in the scene. The use of an object-oriented database is seen as offering the mechanism to create a multilevel geographic model where important contextual relationships including containment can be easily represented. In addition, specific object confirmation or recognition procedures, including the parameter information derived from the first phase, can be attached to the appropriate class of object in the model.

15.8 DISCUSSION

A conceptual model has been presented describing the framework of a system for detecting land-use change between existing map data and high-resolution imagery. Although in the very early stages of the project, an implementation of the first phase is showing the potential to confirm existing map vector objects from imagery. Because of time constraints the approach has made use of mainly 'off-the-shelf', remote sensing and GIS functions and a substantial amount of macro programming, using systems which were available and which could be utilized within the time available. The full implementation of a system, with particular emphasis on the object-model used to store the GIS map data and to 'drive' the system, will require a GIS with tools for raster – vector integration and an object-oriented GIS database. In addition the ability to associate 'expert' recognition functions to particular classes of object is an attractive feature of such systems. With this in mind, the more long-term development of the system will take place using IGIS, the Integrated Geographical Information System, developed by Laser-Scan Ltd.

The main focus of the work has been the detection of change, although a critical assessment of the potential of the techniques for cartographic updating of existing vector maps will be made. At this stage, however, the data being used is considered ill-suited to this purpose, and the implementation of the second phase of the project is required before any clear statements can be made as to whether the proposed methods offer any realistic hope of automatically extracting cartographic – quality features from the imagery. Certainly, other recent studies have shown that, within the context of a National Mapping Agency production flowline (Jamet *et al.*, 1995 and Murray, 1997) much caution is required when considering automated update from imagery.

Early results of the work only really address the first of the three research questions posed at the beginning of this paper, and in addition several clear directions for continued research have emerged.

■ To what extent can we simulate high-resolution satellite data with confidence by degrading digital orthophotos? To what degree do features characteristic of ortho-photos affect the results of integration/matching which may be less important with data from satellite-borne sensors? A subject of current research is the comparison of results from map–image matching for image resolutions from 1 m to 5 m, to establish exactly where the current methods break down.

■ To what extent could the current edge detection procedure be improved? A compara-tive study is necessary to find the procedures which are the most robust at certain pixel resolutions.

■ An assessment of the sensitivity of the vector–edge matching procedure to the type of generalization routine applied to the extracted edges is considered essential.

■ Different styles of vector data, and different levels of structuring, will have a pro-found effect upon what can be achieved by a system such as this. It is argued that the more highly-structured and knowledge-rich the map data is, the more truly top-down the approach can be. An assessment of the different approaches to structuring the vector data, including the benefits of a multi-layered object model for controlling contextual information, will be an important ongoing aspect of this research.

■ Exact mechanisms for implementing the control strategy, and specific object-recognition techniques, may necessitate particular parameters being extracted during the first phase which are not being currently considered. Also, the incorporation of other clues, such as building height automatically derived from stereo images, is being investigated. Research has so far concentrated on panchromatic imagery, but an inter-esting study would be to consider the relative merits of slightly lower resolution, multi-spectral imagery for object extraction, focusing on the specifications of future satellite sensors.

Finally, it is felt that, as the information content of the image increases, and so potentially more useful contextual information becomes available, the increased complexity and volume of data means that the management and control of the change-detection and object-recognition processes becomes more complex. The biggest single challenge of research in this field will be controlling the complexity of the problem as a whole.

15.9 CONCLUSIONS

An approach to automating change detection within a GIS environment has been pro-posed, which uses existing vector maps to guide the recognition of objects in more current imagery, and thereby to detect changes in land use at the object rather than the pixel level. It is felt that this integration of existing map data may allow not only direct comparisons with the image to be made, but also the automatic parameterization of object-recognition modules. The extraction of knowledge from the first phase of this process, in the form of a refined rule-base particular to this image–vector map pair, is seen as a means of 'training' the second phase to improve the recognition of objects which are not represented in the vector data set.

It is believed that research which integrates ideas from GIS, remote sensing and machine vision is crucial for exploring the true potential of very-high-resolution earth-observation data for use in a GIS, both for land-use-change detection and, potentially, for assisting the updating of vector databases.

Acknowledgements

This research work was made possible through University of Nottingham Grant NLRG772. Imagery used courtesy of the National Remote Sensing Centre, Barwell, Leicestershire, UK.

References

BLAMIRE, P.A. and BARNSLEY, M.J. (1996) Inferring Urban Land Use from an analysis of the spatial and morphological characteristics of discrete scene objects, *Proc. 22nd Ann. Conf. Remote Sensing Society*, 11–14 September 1996, Durham.

BOLDT, M., WEISS, R. and RISEMAN, E. (1989) Token-based extraction of straight lines, *IEEE Transactions on Systems, Man and Cybernetics*, Vol. 19, No. 6.

COCHRAN, S.D. (1995) Adaptive vergence for the stereo matching of oblique imagery, *ISPRS J. Photogrammetry and Remote Sensing*, **50** (4), 21–28.

COPE BOWLEY, T. and PIPER, S. (1995) Monitoring rapid urban land-cover change from space, *Proc. sem. on Integrated GIS and High Resolution Satellite Data*, British National Space Centre/Defence Research Agency, 9–10 November 1994, London.

DOWMAN, I. (1997) High Resolution Satellite Data: Boon or Bane, *From Space to Database: Proc. sem. on High Resolution Optical Satellite Missions*, on behalf of the British National Space Centre, 26 February 1997, London.

EVANS, H.F.J. (1996) Neural Network Approach to the classification of urban images, *Unpublished PhD thesis*, Department of Geography, University of Nottingham.

FAUGERAS, O., LAVEAU, S., ROBERT, L., CSURKA, G. and ZELLER, C. (1995) 3-D reconstruction of urban scenes from sequences of images, in A. Gruen, O. Kuebler and P. Agouris (Eds), *Automatic extraction of man-made objects from aerial and space images*, Monte Verita, 145–168, Basel: Birkhauser Verlag.

FORSTNER, W. (1995) Mid-level vision processes for automatic building extraction, in *Automatic Extraction of man-made objects from aerial and space images*, Monte Verita, 179–188, Basel: Birkhauser Verlag.

GRUEN, A. and LI, H. (1995) Road extraction from aerial and satellite images by dynamic programming, *ISPRS J. Photogrammetry and Remote Sensing*, **50** (4), 11–20.

GULCH, E. (1995) Information extraction from digital images: A Kth approach, in multispectral aerial and satellite images, in A. Gruen, O. Kuebler, and P. Agouris (Eds), *Automatic extraction of man-made objects from aerial and space images*, Monte Verita, 63–72, Basel: Birkhauser Verlag.

HARDIE, R.C. and BONCELET, C.G. (1995) Gradient-based edge detection using non-linear edge enhancing pre-filters, *IEEE Transactions on Image Processing*, Vol. 4, No: 11, Nov. 1995.

JAMET, O., DISSARD, O. and AIRAULT, S. (1995) Building extraction from stereo pairs of images: Accuracy and productivity constraint of a topographic production line, in A. Gruen, O. Kuebler and P. Agouris (Eds.), *Automatic extraction of man-made objects from aerial and space images*, Monte Verita, 63–72, Basel: Birkhauser Verlag.

KIM, T. and MULLER, J.P. (1995) Building extraction and verification from spaceborne and aerial imagery using image fusion techniques, in A. Gruen, O. Kuebler, and P. Agouris (Eds), *Automatic extraction of man-made objects from aerial and space images*, Basel: Birkhauser Verlag.

KIM, T. and MULLER, J.P. (1996) Automated urban area building extraction from high resolution stereo imagery, *Image and Vision Computing*, **14**, 115–130.

LASER-SCAN LABORATORIES LTD (1996) *VTRAK Manual*, Laser-Scan Laboratories Ltd, Cambridge Science Park, Milton Road, Cambridge.

LIN, C., HUERTAS, V. and NEVATIA, R. (1995) Detection of buildings from monocular images, in A. Gruen, O. Kuebler and P. Agouris (Eds), *Automatic extraction of man-made objects from aerial and space images*, Basel: Birkhauser Verlag.

MACKWORTH, A.K., MULDER, J.A. and HAVENS, W.S. (1985) Hierarchical arc consistency: exploiting structured domains in constraint satisfaction problems, *J. Computer Intelligence*, Vol. 1.

MAITRE, H., BLOCH, I., MOISSINAC, H. and GOUINAUD, C. (1995) Cooperative use of aerial images and maps for the interpretation of urban scenes, in A. Gruen, O. Kuebler and P. Agouris (Eds), *Automatic extraction of man-made objects from aerial and space images*, Monte Verita, 63–72, Basel: Birkhauser Verlag.

MCKEOWN, D.M., HARVEY, W.A. and MCDERMOTT, J. (1985) Rule-based interpretation of aerial imagery, *IEEE Transactions on Pattern Analysis and Machine Intelligence*, Vol. PAMI-7, No. 5, 570–585.

MCKEOWN, JR. D.M., MCGLONE, C., COCHRAN, S.D., HSIEH, Y.C., ROUX, M. and SHUFELT, J. (1996) Automatic Cartographic Feature Extraction Using Photogrammetric Principles, in C. Greve (Ed.), *Digital Photogrammetry: An Addendum to the Manual of Photogrammetry*, American Society for Photogrammetry and Remote Sensing, Ch. 9, 195–211.

MUELLER, W. and OLSON, J. (1995) Automatic matching of 3-D models to imagery, in multispectral aerial and satellite images, in A. Gruen, O. Kuebler and P. Agouris (Eds), *Automatic extraction of man-made objects from aerial and space images*, Monte Verita, 63–72, Basel: Birkhauser Verlag.

MURRAY, K. (1997) Space Imagery in National Geospatial Database Management, *From Space to Database: Proc. sem. on High Resolution Optical Satellite Missions*, on behalf of the British National Space Centre, 26 February 1997, London.

PLUMB, K. (1997) Review of High Resolution Missions, *From Space to Database: Proc. sem. on High Resolution Optical Satellite Missions*, on behalf of the British National Space Centre, 26 February 1997, London.

PRIESTNALL, G. (1994) *Machine Recognition of Engineering Drawings*, Unpublished PhD Thesis, Department of Computer Science, University of Nottingham.

PRIESTNALL, G., MARSTON, R.E. and ELLIMAN, D.G. (1996) Arrowhead recognition during automated data capture, *Pattern Recognition Letters*, Vol. 17, 277–286.

POULTER, M.A. (1996) On the integration of earth observation data: Defining landscape boundaries to a GIS, in P.A. Burrough and A.U. Frank (Eds), *Geographic objects with indeterminate boundaries*, GISDATA Series, London: Taylor & Francis.

SHETTIGARA, V.K., KEMPINGER, S.G. and AITCHISON, R. (1995) Semi-Automatic detection and extraction of man-made objects in multispectral aerial and satellite images, in A. Gruen, O. Kuebler and P. Agouris (Eds), *Automatic extraction of man-made objects from aerial and space images*, Monte Verita, 63–72, Basel: Birkhauser Verlag.

SMITH, G., FULLER, R., AMABLE, G., COSTA, C. and DEVEREUX, B. (1997) Classification of Environment with vector and raster mapping (CLEVER mapping), *Extended abstracts from the Geographical Information Systems Research – UK Conference (GISRUK 97)*, 9–11 April 1997, University of Leeds, Leeds, UK.

SONKA, M., HLAVAC, V. and BOYLE, R. (1993) *Image Processing, Analysis and Machine Vision*, London: Chapman and Hall Computing.

STILLA, U. (1995) Map-aided structural analysis of aerial images, *ISPRS J. Photogrammetry and Remote Sensing*, **50** (4), 3–10.

STILLA, U., MICHAELSEN, E. and LUTJEN, K. (1995) Structural 3D-Analysis of aerial images with a blackboard-based production system, in A. Gruen, O. Kuebler and P. Agouris (Eds), *Automatic extraction of man-made objects from aerial and space images*, Monte Verita, 53–62, Basel: Birkhauser Verlag.

WEIDNER, U. and FORSTNER, W. (1995) Towards automatic building extraction from high-resolution digital elevation models, *ISPRS J. Photogrammetry and Remote Sensing*, **50** (4), 38–49.

WILKINSON, G.G. (1996) A review of current issues in the integration of GIS and remote sensing data, *Int. J. Geographical Information Systems*, Vol. 10, No. 1, 85–101.

WORBOYS, M.F. (1994) Object-orientated approaches to geo-referenced information, *Int. J. Geographical Information Systems*, Vol. 8, No. 4, 385–399.

WORBOYS, M.F. (1995) *GIS: A Computing Perspective*, London: Taylor & Francis.

ZELLER, H. and DOEMENS, G. (1982) Industrial Applications of Pattern Recognition, *Proc. 6th Int. Conf. on Pattern Recognition*, IEEE Computer Society, 202–213.

PART FIVE

Applications

CHAPTER SIXTEEN

Applying GIS to determine mystery in landscapes

JONATHAN BALDWIN

16.1 INTRODUCTION

GIS has been widely proclaimed as offering an efficient, user-friendly environment for the examination and presentation of digital data. While this may be true in some cases, a notable example being the way geographically-referenced census data can be examined, treatment of more qualitative data categories has proved far less satisfactory. The capabilities of GIS in the quantitative arena has long masked the considerable inadequacies which the technology exhibits when considering the generation and manipulation of less deterministic data. An example of this would be qualitative descriptions of landscape. In addition, it has been accepted for some time that GIS has application shortcomings in another field – that of visual determination. With respect to functions such as viewshed generation and visible area analysis, commentators (Fisher, 1996; De Floriani, 1994) have illustrated inadequacies in the way in which GIS identifies viewshed characteristics and how they are implemented.

16.2 LANDSCAPE QUALITY ASSESSMENT AND GIS

In the wider context of landscape quality assessment, the definitive approach to the satisfactory incorporation of qualitative criteria within GIS appears to have remained elusive. Ribe (1982) stated that there was 'something incongruous about putting a number on scenic beauty', but there still appear to be few alternatives to this in the digital modelling arena. Landscape planning and management depend on the definition of measures of scenic quality, including drama, mystery and interest. Attempts have been made to assess such landscape characteristics through many different methods, but only a few (such as the Lake District National Park) have adopted a GIS approach. These have included the comparison of photographs (Shafer, 1977; Gobster and Chenoweth, 1989) and the application of relative judgements expressed through inventories (Daniel and Boster, 1976) or numeric ratings (Leopold, 1969; Miller, 1956; Linton, 1968). Others have examined the use of video (Bishop, 1995) and photo-realism/montages (Parsons, 1995) while the utilization of purely subjective questionnaire responses to qualitative assessments are employed by others in an attempt to achieve the same goals (Countryside Commission Report, 1986). To date, the majority of acknowledged literary approaches have been based upon the examination of cognitive criteria such as security, risk, and complexity (Berlyne, 1960), as well as addressing variables of texture, depth and measures of the physical landscape (Gobster and Chenoweth, 1989). However, there is at present evidence in the literature to illustrate the incorporation and analysis of such data types within GIS. This paper seeks to refine different facets of these approaches in a digital context, while being principally concerned with the determination of one such measure – that of mystery.

16.3 MYSTERY AND GIS

It is suggested that mystery, where present, plays a significant part in a visitor's choice of landscape experience. Previous assessments of mystery as a component of visual resource management and landscape preference have been undertaken by authors such as Kaplan and Kaplan (1982), Ulrich (1977) and Berlyne (1960). Mystery 'involves the feeling that there will be more to learn, that the communication points to new possibilities still to be explored' (Kaplan and Kaplan, 1982). Mystery is closely linked to the distance that a viewer can see and the shape of the visible area and, in the context of landscape management, can be interpreted as being closely related to the concept of discovery (Baldwin et al., 1996). Mystery can also be used as an indicator of people's behaviour at viewpoints, being a useful component in measuring a visitor's satisfaction with the landscape they are observing. It can also be seen as a determining factor in the extent to which an observer of the landscape becomes a participant. In this way mystery forms a small but significant stepping-stone in the cognitive analysis of the landscape, as a viewer appraises the worth of leaving a place of observation and becoming more interactive with the landscape in view. This interaction can be seen as any involvement or activity within the landscape that requires leaving a former place of security – i.e. a car, a viewpoint, etc.

This paper attempts to show how mystery may be determined from GIS functionality in an effort to support site selection within a landscape-planning environment. If mystery can be qualified through the interrogation of digital data based upon certain broad user parameters, it could be used to complement existing policy considerations that are included in the management of such areas.

16.4 MYSTERY AND QUESTIONNAIRE STRUCTURE

Unlike some previous attempts at associating subjective criteria with landscape preference, this paper does not suggest that photographic representations of scenes should be used for analysis. A combination of four site visits with 200 questionnaire surveys was undertaken to establish a user response to the landscape as a basis for the investigation. The questionnaire was designed to identify a broad range of components that were considered contributory factors in the generation of mystery. Assessment of the features within the landscape was introduced through non-directive questions, allowing the respondent to interpret the landscape in view without suggestive terminology or preconceptions being introduced by the interviewer. Of additional consideration were the previously suggested elements of familiarity and knowledge (Purcell, 1992), focality (Ulrich, 1979) and interest (Baldwin et al., 1996). The focus of the questions was the analysis of landscape characteristics visible from a single vantage point. The final question asks for a binary distinction as to whether the landscape in view was considered mysterious.

The structure of the questionnaire was based around determining the following criteria:

■ *Identify* the principle nature and number of previous visits.

■ *Assess* the perceived impact of landscape features on visitors.

■ *Identify* which features are considered attractors and which detractors.

■ *Examine* the dominance of routeways (paths, tracks etc.).

■ *Assess* the relative impact of horizons, viewshape and distance components.

■ *Identify* the preferred viewing location within the given landscape.

■ *Consider* the impact of land surface features, human impacts and landcover.

■ *Collate* personal information: age cohort, gender, occupation etc.

It is proposed that the above information can be incorporated within and combined with GIS functionality in an attempt to replicate the ways in which people may respond to mystery in the landscape.

16.5 GIS-BASED APPROACH

It is the author's suggestion that for GIS to continue to be considered as an effective tool in environmental management, existing technology needs to find ways to address the more qualitative ideas that typify user behaviour. Once this is achieved, these components can be integrated as part of policy and decision-making frameworks. Of particular interest to this author is the identification of the relationships that may exist between the physical characteristics present in a scene and the ways in which observers respond to them in the wider landscape context. Whilst it appeared relatively easy for respondents at the viewing locations to identify different landscape elements and then place them into categories, it is not yet possible for a GIS to do the same in an intelligent any from digital information.

Combining the information obtained and building upon established parameters of the viewshed function, steps were taken to generate a non-computer-intensive way of identifying the specific criteria that characterize the viewshed in question. Of particular importance were a) the number and type of horizons in the view, b) the viewshadow and c) the shape of the view.

16.5.1 The horizons

The horizons within each view were sorted into three principle group-types – the proximal, intermediate and skyline horizons. This was done by combining the total number of horizons between the viewer and the skyline with the distance of the horizons from the viewing point (Table 16.1).

Table 16.1 Identification of horizon type.

Horizon type	Distance from viewer	Position and Appearance
Proximal/Local	Up to 50–500 m	Closest and clearest
Intermediate	500 m–5 km	Middle distance and quite clear
Distal	5 km +	Far distance and Hazy

The length of the horizon features within the given viewshed may be calculated to give an indication of horizon-definition within the visible area. Kaplan (1982) suggests that mystery depends on a strong element of continuity particularly if the settings 'imply that the new information will be continuous with, and related to, that which has gone before'. Therefore, it is anticipated that proximal horizons will generate an increased fear of the unknown, since what is ahead cannot be related to what has been previously experienced. Skyline horizons will tend to generate interest rather than mystery as the continuity aspect is broken by distance, while intermediate horizons, between 1 and 5 km, may be interpreted as mysterious, particularly if open ground or a path is visible beyond it. Once identified, these horizons can be linked to the viewshadow.

16.5.2   The viewshadow

In this context, the viewshadow is taken to be the sum of all the areas masked by the horizons in the viewshed, along the line of sight from the viewing point to the skyline. The viewshadow (also known as intervisibility or dead ground) is based upon the ratio of visible to invisible ground within the envelope of the viewshed. Closely allied to vegetation and landcover type, this component was found to mirror responses of perceived mystery. An increase in mystery was matched by an increase in the viewshadow. Vegetation transparency and the distance to the access point or transition between visible and invisible zones was significant in determining security. Combined with the horizon component, it is suggested that mystery can also be transposed into knowledge or fear, depending on the kind of landscape characteristics present in the view.

16.5.3   The viewshape

The shape of the view is considered important when determining view-type classification. The viewshape component suggested in this research builds on the graphic landscape typology suggested by Craik (1972) and work by Miller *et al.* (1994). The maximum distance visible in view, relief angles (the angles at which the observed horizons intersect as seen from the viewpoint) and the characteristics of the skyline and horizons are used to categorize the landscape into four viewshape categories. These 'shape categories' are pre-selected as controls for the comparisons of area and linear features between given viewsheds.

Table 16.2   Definition of viewshapes.

Viewshape	Characteristics
Panoramic	>180° vision of the landscape; no dominant intermediate horizons
Vista	a view that is unseen from all approach paths and is only visible when observer arrives at the viewpoint itself. Often bounded locally by elements such as vegetation or physical features
Corridor	a view that allows partial visibility into an area beyond a horizon, the presence of which prevents full disclosure of the landscape beyond. Channelled focus of the observer.
Framed	a view typified by sloping horizons on each side, not quite meeting in the middle, and thus providing a framing effect on the area between them.

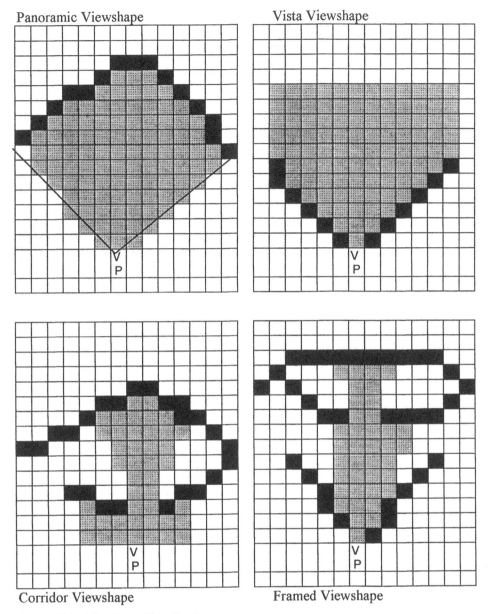

Figure 16.1 Viewshape identification.

The *four* principle viewshapes considered here are:

1 Panoramic.

2 Vista.

3 Corridor.

4 Framed.

The identification of the viewshapes from the digital data may be achieved by classifying the landscape according to the criteria considered in Table 16.2. The anticipated results are shown in Figure 16.1.

16.6 IDENTIFICATION OF THE VIEWSHAPE

The following representations of the viewshapes can be generated using a TCL/TK-based program, which simplifies the procedures adopted by a conventional viewshed algorithm. By using height values from a given viewing point for each pixel in view, an impression of the visible area can be generated. It is accepted that the generation of viewshape in this manner is done so on a relatively coarse scale, but the broad viewshape is clearly interpretable. The shapes are given as plan drawings, with the black pixels representing horizons and the grey pixels, area in view. VP represents the viewing position.

The ratio of the areas in view and out of view can then be calculated to compare viewshape and viewshadow. Comparison of the three-dimensional and planametric characteristics may then be undertaken with consideration being given to the elements discussed above.

In addition to the above components considered, the vegetation characteristics of the analysis area must be included (Shafer, 1969) as well as the evidence of human impact on the landscape. Human impact is widely considered as having a direct effect on cognitive responses to the landscape, and acts as a determining factor in concepts such as security and risk, and consequently, mystery. For example, observers are more likely to feel secure in a landscape given a visible presence of other people. A more predominant level of human activity is likely to reduce the perceived mystery of a scene. Whilst there are many landscape features that can be considered a function of human impact, *three* general classes are used here to represent the effect on the landscape surface. *Routeways* are interpreted as a paths, bridleways, and metalled roads; *buildings* include dwellings and non-residential structures; and *cultural features* are elements of historical and archaeological interest. The horizon characteristics, viewshadow, viewshape and the landcover data can be combined through a series of GIS procedures to produce a means by which the results from the questionnaire survey can be interpreted. This process is illustrated in Figure 16.2.

There are fourteen possible components that comprise the GIS assessment of the mystery rating. Each step of the analysis generates a sub-result of the level of mystery in the view, and these are combined in the final section of the digital data interrogation to generate a mystery rating. The rating is on a scale of 0–12, with 0 representing a low mystery rating and 12 a high one.

Initial indications from the qualitative data are as follows:

- A landscape that an observer looks up or into is likely to be more mysterious than one looked down on. This is because to be at the top, one has had to traverse the intermediate territory, thus conquering the unknown, reducing mystery and making it safe.
- Mystery is closely linked to knowledge and interest.
- The more visits made by an observer to the same viewing location, the more likely it is that mystery will be superseded by interest as the principle component in the scenic assessment.
- A corridor view is likely to be more mysterious than a panoramic one.
- A viewshed with an equal ratio viewshadow to area in view is likely to be seen as mysterious. The greater the percentage of viewshadow in the scene, the more likely is the experience of fear.
- A medium level of mystery is the most likely to attract involvement. A low mystery element would not be likely to stimulate, and a high mystery element would verge on fear and discourage interaction.

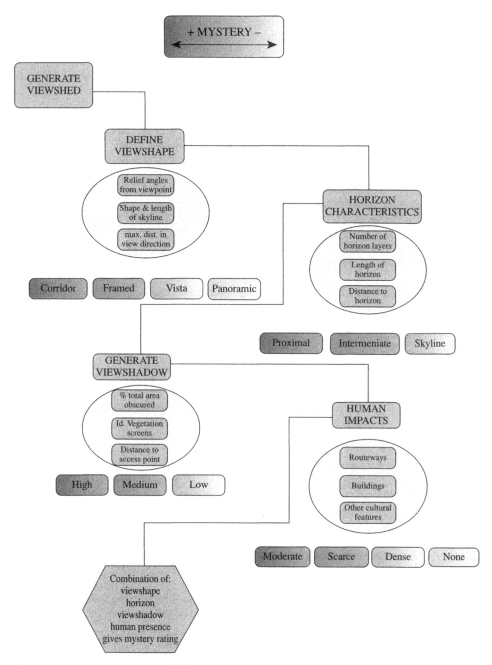

Figure 16.2 GIS assessment of the mystery rating.

The author proposes that, through GIS applications and the identification of key cognitive components, it will be possible to substantiate the above indications, and provide an optimum combination of factors within any given landscape that suggest the degree to which mystery is present in the assessment of landscape quality. It is the intention that further research in the GIS arena will explore other cognitive criteria in observer–landscape relationships that are currently given notional consideration in landscape management and quality assessment. This would provide a means by which the gap between established viewshed algorithms and accepted subjective criteria in respect of landscape preference and evaluation may be bridged.

References

BALDWIN, J.A.D., FISHER P.F., WOOD, J. and LANGFORD, M. (1996) Modelling environmental cognition of the View with Geographical Information Systems. NCGIA Conference on Environmental Modelling (Proceedings), New Mexico, Santa Fe, 1996.

BERLYNE, D.E. (1954) An experimental study of human curiosity. *British Journal of Psychology*, **45**, 256.

BISHOP, I.D. (1995) Comparing regression and neural net based approaches to modelling scenic beauty. *Landscape and Urban planning*, **34** (1996), 125–134.

COUNTRYSIDE COMMISSION (1986) Wildlife & Countryside Act 1986; Conservation maps of National Parks: guidelines by CC: Document no. 6.

CRAIK, K.H. (1972) Appraising the objectivity of landscape dimensions. Natural environments: studies in theoretical and applied analysis. edit. Krutilla JV Pub. Resources for the future inc. 292–308.

DANIEL, T.C. and BOSTER, R.S. (1976) Measuring Landscape Esthetics: The SBE Method. USDA Forest Service, Research paper, RM-167.

DE FLORIANI, L., MARZANO, P. and PUPPO, E. (1994a) Line-of-sight communication on terrain models. *International Journal of Geographical Information Systems*, **8**, 329–342.

FISHER, P.F. (1996) Extending the applicability of Viewsheds in Landscape Planning. *Photogrammetric Engineering and Remote Sensing*, **62** (11), 1297–1302.

GOBSTER, P.H. and CHENOWETH, R.E. (1989) The dimensions of Aesthetic Preference: A Quantitative Analysis. *Journal of Environmental Management*, **29**, 47–72.

KAPLAN, S. and KAPLAN, R. (1982) Cognition and Environment: Functioning in an uncertain world. Praeger Publishers, New York.

LEOPOLD, L.B. (1969) Landscape esthetics: how to quantify the scenics of a river valley. *Nat. Hist.*, **78**, 36–45.

LINTON, D.L. (1968) The assessment of scenery as a natural resource. *Scottish Geographical Magazine*. 84, 219–238.

MILLER, D., MORRICE, J., HORNE, P. and ASPINALL, R. (1994) The use of GIS for the analysis of scenery in the Cairngorm Mts. in Heywood, I. & Price, M. (Eds), GIS in Mountainous Regions, London: Taylor & Francis, 119–131.

MILLER, D.R. (1995) Categorisation of terrain views. Innovations in GIS [2]; Selected papers from the IInd National Conference on GIS Research UK, pp. 215–222, London: Taylor & Francis.

MILLER, G.A. (1956) The magical number seven, plus or minus two: some limits in our capacity for processing information. *Psychological Review*, **63**, 81.

PARSONS, E. (1995) GIS visualisation tools for qualitative spatial information. Innovations in GIS [2]; Selected papers from the IInd National Conference on GIS Research UK, pp. 201–212, London: Taylor & Francis.

PURCELL, A.T. (1992) Abstract and Specific Physical Attributes and the Experience of Landscape; Journal of *Environmental Management*, **34**, 159–177.

RIBE, R.G. (1982) On the possibilty of quantifying scenic beauty – a response. *Landscape Planning*, **9**, 61–75.

SHAFER, E.L., HAMILTON, J.F. and SCHMIDT, E.A. (1969) Natural Landscape Preferences: A predictive model. *Journal of Leisure Research*, **1** (1), Winter, 1969.

SHAFER, E.L. and BRUSH, R.O. (1977) How to measure preferences for photographs of natural landscapes. *Landscape Planning*, **4**, 237–256.

ULRICH, R.S. (1979) Visual Landscapes and psychological well-being. *Journal of Landscape Research*, **4** (1), 17.

CHAPTER SEVENTEEN

Habitat suitability analysis using logistic regression and GIS to outline potential areas for conservation of the Grey Wolf (*Canis lupus*)

NOEMI DE LA VILLE, STEVEN H. COUSINS AND CHRIS BIRD

17.1 INTRODUCTION

Distribution patterns of large carnivores are strongly influenced by environmental discontinuity, human persecutions and other human activities. Their distribution can be regulated by extrinsic factors such as: weather conditions, food supply, vegetation and human disturbance in the landscape (Schamberger and O'Neil, 1986; Yalden, 1993). Understanding why similar regions show different distribution patterns could also be explained by historical reasons of animal–human conflicts. Large-bodied predators, particularly the Grey Wolf, have been persecuted for animal damage control (Fritts and Mech, 1981), due to their economic value – e.g. fur – and out of fear (Boitani, 1996). Moreover, the fact that in terrestrial systems, large carnivores are principally limited to a few species of mammals occurring in low density populations[1] makes this group more susceptible to human perturbation (Yalden, 1993).

Modelling the relationships between animals and their biotic and physical environment has been used in planning studies and it provides a framework around which habitat information can be structured for decision-making. These animal–habitat interactions provide a consistent basis for impact assessment, mitigation, baseline, conservation and monitoring studies (USFWS, 1980; Lancia *et al.*, 1986; Morrison *et al.*, 1992). Such models, based on the concept of habitat, which is defined as the ability of the habitat to provide life requisites (Schamberger and O'Neil, 1986), may yield biological insight to explain the distribution of wolves and evaluate land-management activities.

In this context, the aim of the paper is to analyze the distribution of the Grey Wolf in the Nearctic zoological realm by means of GIS and multivariate statistical modelling. We first predict the probability of wolf-occurrence used as a proxy of habitat quality.

Secondly, the model is validated on an independent set of samples and classification error rates are estimated. Finally, the overall model output in the form of a probability map is used for selecting wolf-habitat types. These are compared with information regarding protected areas to allocate potential regions for conservation or reintroduction. The approach followed here is practical, allowing one to make statistical inferences while yielding biological insight, and it summarizes and predicts the occurrence of a single species used as surrogate of habitat quality. The results are placed within the context of human impact on a habitat-endangered species with the emphasis on predator–human relationships.

17.2 BACKGROUND

17.2.1 Development of a conceptual framework

It has been proposed (Cousins, 1990) that the larger ecological unit that can be made by ecological processes is the food web of the top predator. Indeed, it is suggested that this food web structure is the physical representation of the ecosystem object as opposed to the traditional ecosystem concept of Lindeman (1942). The top predator's territory defines the limits of these ecological units which include all organisms found within its territorial boundary. In a system with no human intervention, the top predator of the food chain is normally the largest species that can predate the largest prey species. They have prey but not predators (Brian and Cohen, 1984). These units have been named by Cousins (1990) 'Ecosystem Tropic Modules' (ETMs). The ETM is a countable object that can be mapped since it has spatial attributes. More broadly the ETMs can be seen as units which retain the food web dynamics and biodiversity of a whole ecosystem.

Here we propose that modelling the relationships of environmental attributes and the size of the ETM could be used to define potentially suitable areas for protection with different levels of human intervention. For this study we made some simplifying assumptions when building the wolf-habitat suitability model:

1 The probability of wolf presence was used as a proxy of habitat quality. It is assumed that there is a highest probability of finding the species in areas that satisfy the species' life requirements (Fritts and Mech, 1981).

2 Wolf packs have non-overlapping home range (Fritts and Mech, 1981), and they use it evenly.

3 Similar home range size was assumed for different habitat types.

4 The geographical range of the species is spatially discontinuous. Home ranges are not uniformly distributed in space, but rather they tend to be clustered together in regions of suitable habitat (Fritts and Mech, 1981). Thus the geographical range is viewed as clusters of ETM defined by wolf metapopulations.

17.2.2 Study area and distribution of the species

The study area was defined by the former wolf geographical range over the Nearctic realm according to the Udvardy classification (Udvardy, 1975). It comprises Greenland and North America without the southern tip of Florida (the Everglades). In this realm the wolf was found, except for the southern part of the USA, where the Red Wolf (*Canis rufus*) was present. The present wolf range has been considerably reduced, mainly in the conterminous United States and Mexico (Nowak, 1981).

The Grey Wolf has the greatest natural range of any living terrestrial mammal other than man. It was found in almost all habitats of the northern hemisphere except equatorial tropical forests. Wolves are habitat generalists. They occupy habitats such as: tundra, taiga, subalpine, mixed and hardwood forests, steppes, grasslands, chaparral and deserts (Nowak, 1981). The study of the species and its habitat has also been undertaken in the Palearctic and Indo-Malayan realms (Europe, Asia and the Arabian peninsula) for comprehension of the wolf-habitat relationships over its global geographical range (De La Ville, 1997).

17.3 METHODS AND PROCEDURE

17.3.1 Construction of the digital database

The selection of the variables and their assemblage into the GIS are the most expensive and time-consuming steps for spatial analysis. They generally involve tailoring databases of different storage format, quality and resolution. The choice is governed largely by the available budget and the purpose of the study. Despite these constraints, GIS users must be sure that the resulting database is as free as possible from error. The GIS software application used here was ARC/INFO version 7.0.4 on a UNIX platform for storage, editing and handling the GIS layers, and the PC version 3.4 for digitizing (ESRI, 1994).

Source materials of the wolf distribution included national and regional maps published in journal and literature references (for a review showing the sources used, see De La Ville, 1997). It also included information from the IUCN-SSC Wolf Specialist Group and the International Wolf Center. The source for the habitat-descriptors included digital information in map form with different format types. Some of the sources were themselves compilations and others were based on source data. The most reliable and most recent source material was used, but in some instances older material was more accurate than the newer sources. The overall result was a hybrid database which emphasized spatial information rather than temporal uniformity.

A set of nine habitat variables were selected and layers were built using a series of GIS operations. The dependent variable was recorded in binary format as wolf-presence or absence. The borders of the predator distribution were digitized at 1:1 000 000 scale using the Digital Chart of the World (DCW) as contour maps (ESRI, 1991). Information on elevation and road networks were also obtained from the DCW. The unequal interval classes for elevation were transformed into a distribution frequency for statistical analysis in order to avoid using this attribute as a dummy variable. From the road network a layer data regarding road density (km/sq. km) was estimated using GIS capabilities.

Information regarding population of cities over 200 000 inhabitants was collected (UNEP, 1985) and transformed into a sphere of influence according to population size. Buffers were generated around each point defining the distance according to the Central Place theory (Christaller, 1930 in Waugh, 1995). The buffer's area varies in size from a small village to a primate city and forms a link in urban hierarchy.

The land-cover data is the Wilson and Henderson-Sellers classification (1985) recorded on two separate arrays as primary and secondary land-cover types. The original classification scheme included 53 vegetation classes based on physiognomy, density and seasonal variation. The classes were broken down into 17 components and each cell was assigned the absolute percentage of cover for modelling purposes. The components were then regrouped on the basis of physiognomy and agricultural areas, for a total of four categories: percentage of cover for *woody* (trees and shrubs) and *herbaceous* (grass, crop and tundra) elements in *natural* and *modified* areas. The original classification scheme was also used for generating the land-cover types within the wolf-suitable habitat.

For climatic characterization the data set used was the IIASA database. It provides a global terrestrial grid of 0.5×0.5 degree lat./long compiled from weather records from different sources (Leemans and Cramer, 1991). Total annual rainfall and the average maximum temperature were used as climatic variables.

A digital dataset of protected areas was obtained from the World Conservation Monitoring Centre (WCMC) in ARC/INFO format with region topology. This attribute was not included in the model. It was used to compare the wolf-suitability map and land-cover types with current areas under management. This dataset is classified according to IUCN (1994) into six categories according to primary management objectives.

The analysis was carried out in raster format with a cell size of 10×10 km. This is a trade-off between the average wolf pack territory[2] which defines the size of the ETMs, the original resolution of the data set and CPU-time. All layers and grids were analyzed using the Lambert Azimuthal equal area projection. A cell size of 100 sq. km generated a bounding rectangle of 810 rows and 1 540 columns for a total of 1 247 400 cells, of which 184 640 were contained within the study area.

17.3.2 Analytical design

The nine GIS layers (covariates) and the wolf-distribution layer (dependent variable) were subjected to the fieldwork operation through systematic sampling in ARC/INFO. It was assumed that only one potential observation per location was available. Thus the amount of replication at any site was not considered here as a design parameter. The sample data can be depicted as two-way table or matrix with observations at each location (rows) and variables (layers) or cell descriptors as columns. Systematic sampling schemes were carried out, and spatial autocorrelation using Moran's I coefficient was tested for the third, fifth and seventh-order lag (30, 50 and 70 km). The data matrix selected was input into SPSS version 6.1 for statistical analysis (SPSS, 1994).

One of the objectives of this study was to test the relationship between a dichotomous dependent variable (presence/absence) and independent environmental variables in order to build a model that would best describe in terms of probability the quality of the wolf habitat. Two related, and yet conceptually quite different, techniques could have been applied: logistic regression and discriminant analysis. However, the non-linear logistic regression was selected because the independent variables did not satisfy the assumptions of multivariate normality and equal covariance–variance matrix necessary for discriminant analysis (Cox and Snell, 1970). Logistic regression has been used previously for analyzing anthropogenic deforestation in Honduras (Ludeke *et al*., 1990), modelling kangaroo presence/absence data (Walker, 1990), the distribution of the Mt. Graham red squirrel (Pereira and Itami, 1991) and interpreting bird atlas data (Osborne and Tigar, 1992).

The parameters of the wolf-habitat suitability model were estimated with the maximum likelihood method, using a forward stepwise variable selection. It is important to recognize that this method does not give the best selection of variables in an absolute sense, just the best statistical model that fits the data under a particular set of conditions.

The contribution of each variable to the model was estimated using the partial correlation coefficient (R). The validation of the model output was carried out using a split sample test, where the model is obtained from one part of the dataset and calibrated on another part (Pereira and Itami, 1991). The model equation was estimated on 80% (5 898 cells) of the data and model validation was carried out on the remaining 20% (1 441 cells). The classification error rates were determined from the percentage of correct predictions when the model predicts presence and the wolf is actually absent (error type I) and model prediction on the wolf absence when it was observed as present (error type II). They were estimated by selecting at each cut-off of probability value the number of cells correctly and wrongly predicted with wolf-presence and absence. Changing the classification rules allowed one to group cells to the left of a given point as unsuitable habitat (wolf-absence), while cells located to the right were assigned as suitable (wolf-presence). The classification rule for an optimum cut-off value, for defining suitable or unsuitable habitat, was chosen as the probability with its most successes and lowest failure.

It should be mentioned that areas where wolf-occurrence was unknown (see Figure 17.3a in the colour section), were not included for building the model. Instead these cells were used for model prediction. The model equation was input into ARC/INFO and the result was a map of continuous probability. It was sliced into 10 discrete probability intervals, from 0 to 1.

17.4 RESULTS AND DISCUSSION

17.4.1 Spatial autocorrelation and sampling scheme design

Moran's I autocorrelation index at the third order lag (each 30 km) for each layer was close to unity (~ 0.99). Large values of the coefficient (I > 0) mean that clustered cells have strong spatial similarities on their attributes (Cliff and Ord, 1981). For the fifth and seven order it dropped close to 0, ranging from 0.11 to 0.23. For the seventh order the maximum value was 0.15. The sample cells lagged at fifth and seventh order were independent and uncorrelated. Thus, the systematic sampling design at fifth-order lag, each 50 km, was applied in both directions (Easting and Northing).

17.4.2 Wolf habitat suitability model

The final coefficients and statistics for the best model are summarised in Table 17.1. From the original 9 variables the only one removed from the model was total annual rainfall (its residual chi-square for the coefficient β was greater than 0.05 at 95% confidence level). The rest of the variables met the criterion for remaining in the model, with β differing from 0 at 95% confidence level. Given the coefficients of β in Table 17.1, the logistic model equation for the probability of the event occurring, i.e. suitable habitat, can be written as:

$$\text{Prob}(suitable\_habitat) = \frac{1}{1 + e^{-z}} \tag{17.1}$$

where

$$z = 1.10 - 4.19 \times road\_den - 0.36 \times woody(m) - 0.10 \times hum\_den - 0.06 \times grass(m)$$
$$+ 0.02 \times grass(n) + 0.07 \times woody(n) - 0.05 \times max\_tmp + 0.01 \times elev \tag{17.2}$$

Table 17.1 Final coefficients and statistics for the best model (*m* = modified areas, *n* = natural areas).

Variable	β	Significance level	R
Constant	1.10	–	–
road_den	−4.19	0	−0.13
woody (m)	−0.36	0	−0.13
hum_den	−0.10	0	−0.11
grass (m)	−0.06	0	−0.10
grass (n)	0.02	0	−0.10
woody (n)	0.07	0	0.07
max_tmp	−0.05	0	−0.06
elev	0.01	0.007	0.03

By looking at Table 17.1, it is possible to see that the values of the statistic R are quite similar for the first 7 covariates. This indicates that each variable has a similar partial contribution to the model. However, there is one group positively affecting the probability of the occurring event, while another shows an opposite effect. Variables with positive values for the statistic R indicate that, as they increase, the likelihood of finding a suitable habitat for the species also increases. For negative values, the opposite is true. According to the results shown in Table 17.1, herbaceous elements in natural areas have a positive effect for defining suitable habitat for wolves. This result would seem paradoxical, since open landscapes, i.e. tundra and steppes, offer low protection for wolves from man – e.g. hunting from aircraft (Bibikov, 1994). However, it should be noted that this part of the study area is the least perturbed by human activities. On the contrary, as was expected, the presence of woody elements in natural areas is likely to increase the likelihood of finding suitable habitat, since wolves are less vulnerable in forests and woodlands.

The occurrence of extremely low maximum temperatures, on the other hand, is a factor that determines low habitat-quality and therefore limits the probability of finding wolves (Marquard-Petersen, 1986). The elevation variable has a positive and the smallest contribution to the model output. The highest value for this variable corresponds to the lowest elevation class (0–340 m.a.s.l.). Therefore, high-quality habitat can also be defined by low elevation ranges.

Descriptors used as a measure of human impact (road and human density, woody and herbaceous elements in modified areas) have a negative effect on the species' habitat quality. Human activities, i.e. industrial, agricultural and residential development, tend to modify natural areas, and therefore pose a threat to wildlife and their habitats. Although some areas in North America retain significant and pristine natural habitats, these stocks are nothing like as large as they once were.

The model classification for observed and predicted data on the set of training cells gave an overall percentage of correct prediction of 91.35% at 0.5 probability value. Cells with values assigned as non-suitable habitat (probability of the event not occurring) were correctly predicted with less exactitude (83.23%) than cells allocated as suitable habitat (96.74%). These results give a percentage of error type I of 16.77% (cells predicted with

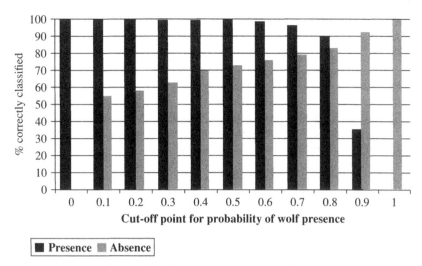

Figure 17.1 Model validation: classification error rates on the 20% of the original sample data in the Nearctic realm (NEA).

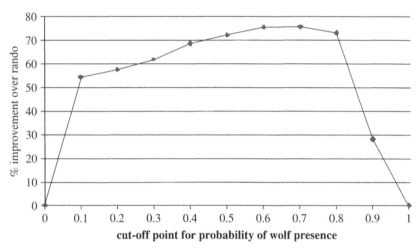

Figure 17.2 Improvement over a random set of samples for classification rules at different cut-off probability values.

wolf-presence when it is actually absent), and error type II of 3.26% (cells predicted with wolf-absence when it is actually present).

Figures 17.1 and 17.2 display the results of the classification errors rates and improvement over a random set of samples (20% of the original sample data) at different cut-off probability values. From Figure 17.1 it is possible to estimate that, for a cut-off value of 0.6, the model correctly predicts 98.51% of the cells for wolf-presence with just 24.17% of the cells wrongly classified with wolf-presence when in fact the species is absent (error type I). Subtracting these values, an improvement of 75.34% of cells correctly classified at 0.6 probability value was achieved (Figure 17.2). This is the optimal cut-off value at which the model performs with the most predictive success and lowest failure. Values over 0.6 therefore will be considered as indicative of a suitable habitat for wolves, and below that point an unsuitable one.

According to the model output, suitable areas are characterized by: low human perturbation levels (road density <0.1 km /sq. km, coverage of agricultural areas <10% and human density <5 inhabitants/sq. km); percentage cover by herbaceous and woody elements >40%; maximum temperatures below 14°C and elevation below 2 000 m.a.s.l.

The fact that the presence of roads on the landscape jeopardizes wildlife has been well documented for some species. In Wisconsin, Thiel (1995) found that road densities over 0.59 km/sq. km affected wolf breeding patterns. Fuller *et al.* (1992) reported similar results in some studies carried out in Minnesota. They reported a threshold of 0.70 km/ sq. km for defining suitable habitat. Secure habitat is reduced for large mammals other than wolves, such as elks, wolverines, lynx and bears, when road densities are over 0.6 km/sq. km. Havlick *et al.* (1996) in a study carried out in the Northern Rockies found that a 0.48 km buffer distance is the minimum distance below which grizzly bears' use of the habitat decreases significantly. They suggested direct and indirect effects of road perturbation on habitats for wildlife. Among the direct effects are: forest fragmentation and increasing human activities, e.g. legal and illegal hunting, increased transport and industrial use of areas with a dense road network. On the other hand, the presence of roads, whether they are used or not, increases peak runoff levels and storm discharge due to local soil erosion.

17.4.3   Selection of areas for conservation or reintroduction

Figure 17.3a (see colour section) displays the model output of suitable areas with the current wolf distribution in the study area. Observed and predicted data for wolf presence and absence agree quite well in Canada and most of the USA. However, there are some regions in Canada, i.e. British Columbia and New Brunswick, and the conterminous United States, i.e. Utah, Colorado, Oregon, California and West Virginia, where wolves are absent and the model predicted them as suitable areas (see Figure 17.3a). Moreover, some zones with an unknown wolf occurrence were predicted as suitable areas, (these cells were not input as primary data for building the model), i.e. Labrador in Canada, the Northern Rockies and the Sierra Madre Occidental in Mexico. These regions have been highlighted in Figure 17.3b, showing their land-cover types. They are as follow: The Northern Rockies in Idaho, Wyoming, Montana (1) and British Columbia and Alberta (2); Labrador (3); New Brunswick and Nova Scotia (4); Minnesota, Michigan and Wisconsin (5); part of the Appalachian mountains in West Virginia (6); Utah, New Mexico and Colorado (7); Oregon and California (8) and the Sierra Madre Occidental in Chihuahua and Durango (9). These areas exhibit different wolf-habitat types (Figure 17.3b), and therefore the wolf should be protected or reintroduced in an example of each major vegetation type. At present there is a bias of wolf-management in protected areas (PA) on Tundra and Taiga vegetation types.

The US Fish and Wildlife Service (1980, 1987) has suggested that the optimum habitat for wolves would have to satisfy the requirements of: (1) A sufficient prey base; (2) suitable, secluded denning and rendezvous sites; (3) sufficient space with minimal exposure to humans; (4) maximum 10% of private land ownership; and (5) absence, if possible, of livestock grazing. Mech (1979) has estimated a minimum area of 10 000 sq. km to assure viable, well-functioning and well-organized wolf populations. He pointed out that if a smaller area is used, human/wolf interactions could be quite high around the edges. An area of 10 000 sq. km could support around 100 individuals (~10 breeding pairs for reintroduction). The Northern Rockies Recovery Plan followed these criteria and the proposed areas were located in North-western Montana, central Idaho and the Greater Yellowstone area (see zone 1 in Figure 17.3b) (USFWS, 1987). In 1995, 14 wolves were released in the Yellowstone National Park (YNP) and 15 in central Idaho. The following year 20 additional wolves were released in central Idaho and 17 in acclimatization pens in YNP (Fritts *et al.*, 1996). They pointed out that, by February 1996, at least 140 wolves were known to exist in the Northern Rockies: >70 in North-western Montana; 37 in the Yellowstone area (including 17 in pens awaiting release); and 33 in South-western Montana.

Combining criteria used by the USFWS for the recovery plan in the Northern Rockies and the minimum area suggested by Mech of 10 000 sq. km, it is possible to select areas for conservation where the wolf is still present or where it could be reintroduced. Considering that one ETM could be defined by the home range of one breeding pair and pups with a maximum number of 10 individuals per pack, the area suggested by Mech gives a conservative estimate of 10 ETMs as the minimum number of units to secure the top predator's population and species under its food chain.

Though zones located in optimal habitat for the Grey Wolf in Alaska comprise large extensions of federally-managed lands and wilderness areas currently under protection (greater than 10 000 sq. km), they include few different wolf habitat types (Figure 17.3b). On the contrary, some suitable areas located in Canada (zones 3 and 4), the conterminous United States (5, 6, 7 and 8) and in Mexico (9) comprise more variety of wolf-habitat types, i.e. deciduous broadleaf forests and woodlands, deciduous shrublands and drought deciduous woodlands, broadleaf shrublands and woodlands. Therefore, in order to secure the top predator's food chain, maintain high biodiversity and manage ecosystems efficiently, a wider range of wolf-habitat types should be considered for establishing new land use zones for wildlife management at the community level.

Finally, it has to be considered that the selection of areas for protection or reintroduction should take into account that wolves are a highly mobile species, which tends to follow prey migration. Recovery zones, therefore, or areas for protection should consider wolf dispersal corridors (Mech, 1996) and land use zoning (Clarson, 1996). These factors are quite important, since suitable areas, with nearly the minimum size to assure viable populations, are usually fragmented into smaller units and surrounded by zones with strong human activities and human presence. Therefore, considering public acceptance and sentiments, wolf corridors and land-use zoning based on existing or reintroduced wolf populations, e.g. setting boundaries for management zones with different levels of wolf control, public and private agricultural lands, will lead to an effective Wolf-management plan.

17.5 CONCLUSIONS

Establishing areas for protection is a controversial issue in wildlife conservation. Moreover, when protection involves predators, diverse and polarized opinions are raised among various public groups. Economic, political, biological, ethical and cultural values are involved in managing areas and protecting large carnivores. Hence, well-designed and well-conducted studies and the fullest support and understanding of people living in the area are needed when establishing land-use zones for wolf-management. Monitoring the whole wolf-geographical range in the Nearctic realm alone is an arduous and expensive task in terms of time and money. Therefore using the ETM with the aid of GIS and modelling could provide baseline studies to outline potential areas for predator-management and wildlife conservation. The outcome derived from this study should be seen as a method of proposing standards for conservation. Assessment of the application of these standards will require ground validation over a suitable timescale.

Acknowledgements

We would like to thank to The World Conservation Monitoring Centre (WCMC) for its contribution in supplying the Protected Areas database and for allowing us access to the 'Wolf data files'. Thanks are also extended to the IUCN-SSC Wolf Specialist group for giving us vital information on wolf distribution and status. Thanks also to Peter Hiett for correcting this manuscript.

Notes

1 'The Pyramid of Numbers' (Elton, 1927).
2 Pack territory size ranges from 40 to 664 sq. km but averaged between 100 to 250 sq. km (Fritts and Mech, 1981).

References

BIBIKOV, D. (1980) Wolves in the USSR, *Natural History*, **89** (6), 58–63.
BOITANI, L. (1996) Ecological and cultural diversities in the evolution of wolf-human relationships, in L. Carbyn, S. Fritts and D. Seip (Eds), *Ecology and Conservation of Wolves in a Changing world*, Canadian Circumpolar Institute, University of Alberta, Edmonton, Alberta, Canada.
BRIAN, F. and COHEN J. (1984) Environmental correlates of food chain length, *Science*, **238**, 956–960.
CLARKSON, P. (1996) Recommendations for more effective wolf management, in L. Carbyn, S. Fritts and D. Seip (Eds), *Ecology and Conservation of Wolves in a Changing World*, Canadian Circumpolar Institute, University of Alberta, Edmonton, Alberta, Canada.
CLIFF, A. and ORD, J. (1981) *Spatial Process: model and applications*, London: Pion Press.
COUSINS, S.H. (1990) Countable ecosystem deriving from a new food web entity, *Oikos*, **57**, 270–275.
COX, D. and SNELL, E. (1970) *Analysis of binary data*, 2nd Edn, London: Chapman & Hall.
DE LA VILLE, N. (1997) The Grey Wolf: Habitat suitability analysis of a top predator over its global geographical range, PhD Thesis, IERC, Cranfield University, UK.
ELTON, C. (1927) *Animal Ecology*, New York: MacMillan.
ESRI (1991) *Final DCW Product Specification*, U.S. Defence Mapping Agency (Project MGGT-0012).
ESRI (1994) *Understanding GIS. The ARC/INFO Method*, Harlow, Essex, UK: Longman Scientific & Technical, Longman Group Ltd.
FRITTS, S. and MECH, D. (1981) Dynamics, movements and feeding ecology of a new protected wolf population in North-western Minnesota, *Wildlife Monographs*, **80**, Louisville, Kentucky: The Wildlife Society.
FRITTS, S., BANGS, E., FONTAINE, J., BREWSTER, W. and GORE, J. (1996) Restoring wolves to the Northern Rocky mountains of the United States, in L. Carbyn, S. Fritts and D. Seip (Eds), *Ecology and Conservation of Wolves in a Changing World*, Canadian Circumpolar Institute, University of Alberta, Edmonton, Alberta, Canada.
FULLER, T., BERG, W., RADDE, G., LENARZ, M. and BLAIR, G. (1992) A history and current estimate of wolf distribution and numbers in Minnesota, *Wild. Soc. Bull*, **20**, 42–55.
HAVLICK, D., STOCKMANN, K. and BECTHOLD, T. (1996) *A regional approach to wildlife conservation & wildland restoration: The roads scholar program*, Proceedings of the Montana Academy of Sciences.

IUCN (1994) *Guidelines for Protected Area Management Categories*, CNPPA with the assistance of WCMC, Gland, Switzerland and Cambridge, UK: IUCN (x + 261pp.).

LANCIA, R.A., ADAMS, D.A. and LUNK, E.M. (1986) Temporal and Spatial Aspects of Species-Habitat Models, in: J. Verner, M. Morrison and J. Ralph (Eds), *Wildlife 2000*, 177–182, Madison, Wisconsin: University of Wisconsin Press.

LINDEMAN, R. (1942) The trophic-dynamic aspect of ecology, *Ecology*, **23** (4), 399–418.

LUDEKE, A., MAGGIO, R. and REID, L. (1990) An analysis of Anthropogenic deforestation using logistic regression and GIS, *J. Env. Mang*, **31**, 247–259.

MECH, D.L. (1970) The wolf: the ecology and behaviour of an endangered species, Garden City, NY: Doubleday/Natural History Press.

MECH, D.L. (1979) Some considerations in re-establishing wolves in the wild, in E. Klinghammer (Ed.), *The behaviour and ecology of wolves*, New York: Garland STPM Press.

MECH, D.L. (1996) What do we know about wolves and what more do we need to learn? in L. Carbyn, S. Fritts and D. Seip (Eds), *Ecology and Conservation of Wolves in a Changing World*, Canadian Circumpolar Institute, University of Alberta, Edmonton, Alberta, Canada.

MORRISON, M., MARCOT, B. and MANNAN, R. (1992) *Wildlife-habitat relationships: concepts and applications*, Madison, Wisconsin: University of Wisconsin Press.

NOWAK, R. (1981) A perspective on the taxonomy of wolves in North America, in L. Carbyn, (Ed.), *Wolves in Canada and Alaska*, Canadian Wildlife Service, Report Series N. 45.

OSBORNE, P.E. and TIGAR, B.J. (1992) Interpreting bird atlas data using logistic models: an example from Lesotho, Southern Africa. *J. App. Ecology*, **29**, 55–62.

PEREIRA, J. and ITAMI, R. (1991) GIS-Based Habitat Modelling Using Logistic Multiple Regression: A Study of the Mt. Graham Red Squirrel, *Photogrammetric Engineering & Remote Sensing*, **57** (11), 1475–1486.

SCHAMBERGER, M. and O'NEIL, L. (1986) Concepts and constrains of Habitat-model testing, in J. Verner, M. Morrison and J. Ralph (Eds), *Wildlife 2000*, 177–182, Madison, Wisconsin: University of Wisconsin Press.

SPSS (1994) *Advanced Statistics*, 6.0, USA: SPSS Inc.

THIEL, R. (1985) Relationships between road densities and wolf habitat suitability in Wisconsin, *The American Midland Naturalist*, **13** (2), 404–407.

UDVARDY, M. (1975) *A classification of the biogeographical provinces on the world*, IUCN Occasional paper, No. 18, Switzerland.

UNEP (1985) Human population in settlement over 200 000, Grid United Nations Environment programme, Nairobi.

U.S. FISH and WILDLIFE SERVICE (1980) Northern Rocky Mountain Wolf recovery plan, Denver, Colorado (67 pp.).

U.S. FISH and WILDLIFE SERVICE (1987) Northern Rocky Mountain Wolf recovery plan, Denver, Colorado (119 pp.).

WALKER, P. (1990) Modelling wildlife distributions using a geographic information system: kangaroos in relation to climate, *J. Biogeo*, **17**, 279–289.

WAUGH, D. (1995) *Geography: an integrated approach*, 2nd edn, 375–380, Walton-on-Thames, Surrey: Thomas Nelson.

WILSON, M.F. and HENDERSON-SELLERS, A. (1985) A Global Archive of Land Cover and Soils Data for use in General Circulation Climate Models, *J. Climatology*, **5**, 119–143.

YALDEN, D. (1993) The problems of reintroducing carnivores, *Symp. Zool. Sco. London*, **65**, 289–306.

CHAPTER EIGHTEEN

Census geography 2001: designed by and for GIS?

DAVID MARTIN

18.1 INTRODUCTION

The 1991 Census of population was the first in the UK to be conducted in what might be called the 'GIS era'. Despite the fact that GIS use in socio-economic applications had been widespread since the mid-1980s (Martin, 1996), the design of a single census geography based on enumeration districts (EDs) for England and Wales was undertaken manually, resulting in the production of a large number of paper maps. Digital boundaries were subsequently created commercially by digitizing maps provided by the Office for Population Censuses and Surveys (OPCS), and the ED91 and ED-line products produced in this way have seen extensive commercial, governmental and academic use both for simple thematic mapping and more complex GIS applications (Barr, 1993; Charlton *et al.*, 1995). The government committee on the handling of geographic information in 1987 strongly endorsed the use of postcodes as the basis for census geography design (DoE, 1987). However, the 1991 geography was based entirely on census-specific EDs, which have no clear relationship with postcodes, except in a very few local authorities, despite the widespread use of postcode geography in other applications (Raper *et al.*, 1992). The manually-designed geography of 1991 continued to incorporate wide variation in ED populations, including the presence of sub-threshold EDs, whose small populations caused them to be merged with neighbouring areas in order to preserve confidentiality. To commentators such as Openshaw (1995), the continued use of enumeration geography for data output, the lack of an integrated digital geography and the general 'neglect of GIS' in 1991 are seen as unacceptable.

In April 1996 OPCS and the Central Statistical Office were merged to form the Office for National Statistics (ONS), which has taken on the role of the census office for England and Wales. This paper describes developments at ONS, and experiments in the use of GIS in census geography design, exploring the potential for a 2001 census geography designed by and for GIS. The rest of the paper is divided into four sections. These cover the method by which census geography was planned before the advent of GIS; the introduction for the 1997 Census Test of a GIS for collection geography planning, and the potential for the development of an entirely new GIS-designed output geography for 2001, quite separate from the geography of data collection. The paper closes with some summary remarks and recommendations for future census geography design.

18.2   CENSUS GEOGRAPHY DESIGN IN THE PRE-GIS ERA

In previous censuses, the ED has formed the basic building block of census geography, which has been the small area geography of both collection and output, although in 1971 small area statistics (SAS) were additionally published for grid squares (CRU/OPCS/ GRO(S), 1980), and 1991 SAS or postcode sectors. EDs are designed to facilitate enumeration, and therefore aim to equalize the workload placed on enumerators. Their design begins by subdividing wards (which are the smallest statutory areas) and parishes, while retaining the ED boundaries used in the previous census as far as possible. Estimates of the area to be covered, numbers of households to be enumerated, language difficulties, and subdivided properties are assembled for each ED from the results of the previous census and contemporary information provided by local authorities. Where statutory boundary changes or population changes result in EDs falling outside acceptable ranges, boundaries are redrawn, trading off population size against difficulty of enumeration, until each ED provides a suitable task for one enumerator. A further constraint is that boundaries are drawn as far as possible along recognizable topographic features to assist enumerators in the field. The resulting EDs inevitably vary widely in geographical size and shape, population size and density, and social composition. These EDs form the basis for geographical aggregation and reporting of census data, and it is therefore unsurprising that output geography bears little relationship to underlying social geography (see, for example, Morphet, 1993).

Separate simultaneous censuses are held in Scotland and Northern Ireland, and it is important to note that the Scottish situation, in particular, is very different from that in England and Wales. In Scotland the extent of each postcode has been identified and mapped by the General Register Office (Scotland) (GRO(S)) since 1973, and constantly updated. Unit postcode boundaries were digitized in 1990. Scottish output areas (OAs) are the smallest possible areas for reporting SAS, because they have been produced by the aggregation of postcode data until they meet confidentiality criteria. There are, therefore, no restricted OAs, change data are more readily calculated and merging census data with postcoded datasets is relatively straightforward. This is the result of non-minimal investment in data, and the systematic use of GIS.

Despite the advent of address-based georeferencing in the mid-1990s, and the technical potential for producing data for ad hoc, user-defined geographies from a secure, intelligent GIS containing fully georeferenced household level census data (Rhind *et al.*, 1990; Rees, 1997) there will inevitably be a need for standard geographies to support the vast majority of census uses in the early 21st Century. How might a GIS-based approach to census-geography design differ from the current practice, and produce more useful outputs? Various authors have set out desirable properties for census output geographies, such as the seven 'tests' provided by Coombes (1995), and reproduced as Table 18.1. Fundamental features of these 'ideal world' approaches are the separation of collection and output geographies, allowing each to be separately optimized; specification of the criteria for such optimization; a call for consistency of approach throughout the UK; and the production of detailed digital geographical products as an integrated part of the census output. Some of these principles are echoed by Charlton *et al.* (1995).

Table 18.1   Seven tests of census geography (Coombes, 1995).

1. Are the building blocks the smallest that the confidentiality restrictions deem to be possible (to allow maximum flexibility of aggregation)?
2. Is each of the areas in a set of building blocks, or other set of areas, defined on a consistent basis across the country?
3. Does this set of areas represent (parts of) 'real-world' entities, such as settlements, which can thus be recognized using these boundaries?
4. Does the set of areas allow comparison with previous census(es) at this level, or for some minimal grouping of areas to create consistent boundaries?
5. Does the set of areas cover the whole of the country without leaving any locations whose data are too sparse to allow them to be published?
6. Are the boundaries of these areas available in digital form?
7. Can these areas be readily and accurately linked by their location coding to the areas used in many non-census datasets?

18.3   GIS FOR COLLECTION GEOGRAPHY DESIGN

A Census Test took place on 15 June 1997 in six areas in England (together with two areas in Scotland and a sample in Northern Ireland, which are not considered here), designed to evaluate a number of innovations in the census form and field procedures (ONS, 1996). A most important development is that the design of the enumeration geography for these areas will have taken place using a custom-written Arc/Info and Oracle GIS application running on Sun Unix workstations at ONS.

This system, known as the Geography Area Planning System (GAPS) uses a combination of 1:10 000 scanned raster mapping; 1991 ED boundaries; information from local authorities; the statutory boundaries present in Boundary-Line, and ADDRESS-POINT counts in order to reassess the grading of each 1991 ED. The majority of these digital databases did not exist at the time of 1991 geography planning. Each ED which falls outside acceptable levels of enumeration difficulty and population size is flagged for re-planning, and the planner is able to interactively redesign the ED on-screen against a raster topographic map background, with dynamic recalculation of all relevant characteristics. A work unit (typically an entire ward) can only be checked back into the central Oracle database when all its constituent parts have been replanned to meet the specified criteria. The system produces detailed maps for each enumerator, and provides address lists showing the addresses to be enumerated in each ED, illustrated in Figures 18.1 and 18.2 respectively. Clearly, enumerators will find some discrepancies between the output of this system and the situation on the ground, but the ability to provide such information at all represents a major advance over previous working practices. At present, postal geography does not form any direct part of the ED-planning process.

The GAPS GIS and data have been used for the empirical work presented here. The use of GIS in this context marks a significant operational change, although the process of ED planning is still essentially that used in previous censuses, as described above. More fundamental is the potential offered for the creation of an entirely separate output geography(-ies), optimized according to quite different criteria, and taking advantage of the power of the GIS for data organization, and developments in automatic zoning procedures (Openshaw and Rao, 1995).

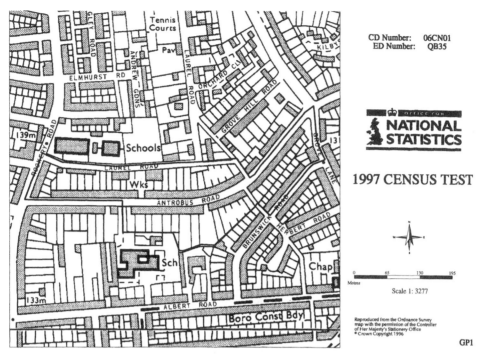

Figure reproduced by kind permission of Ordnance Survey and the Office for National Statistics.
© Crown Copyright. Licence number 87834M.

Figure 18.1 Example ED map output from GAPS.

Part 1

Form No. / Build No.	Address (including Postcode) / (Location or Description of Premises as required)	Advance Round	NR	DEM / Address N/A*	CE	Name of householder or if CE [Manager] [Person in charge] / No. of persons in H/H	DEL	COLL/ REC'D	Dummy Form Information: Type of Accom. / Self contained	No. of rooms / Lowest floor level	Source	*Notes
A	B	C	D	E	F	G	H	I	J	K	L	M
0001	1 ITCHEN ROAD, HIGHFIELD, SOUTHAMPTON, SO17 1BJ											
0002	2 ITCHEN ROAD, HIGHFIELD, SOUTHAMPTON, SO17 1BJ											
0003	3 ITCHEN ROAD, HIGHFIELD, SOUTHAMPTON, SO17 1BJ											
0004	4 ITCHEN ROAD, HIGHFIELD, SOUTHAMPTON, SO17 1BJ											
0005	5 ITCHEN ROAD, HIGHFIELD, SOUTHAMPTON, SO17 1BJ											
0006	6 ITCHEN ROAD, HIGHFIELD, SOUTHAMPTON, SO17 1BK											
0007	7 ITCHEN ROAD, HIGHFIELD, SOUTHAMPTON, SO17 1BK											
0008	8 ITCHEN ROAD, HIGHFIELD, SOUTHAMPTON, SO17 1BK											
0009	9 ITCHEN ROAD, HIGHFIELD, SOUTHAMPTON, SO17 1BK											
0010	10 ITCHEN ROAD, HIGHFIELD, SOUTHAMPTON, SO17 1BK											
	PAGE TOTALS											Page 1 of 14

© Crown Copyright. Figure reproduced by kind permission of the Office for National Statistics.

Figure 18.2 Example address list output from GAPS.

18.4 GIS FOR OUTPUT GEOGRAPHY DESIGN

In the context of census geographies, automated geography design can be considered in two stages. The first of these concerns the construction of suitable 'building blocks', forming small geographically-defined aggregations of census records, while the second concerns the grouping of these building blocks into output areas (OAs) which meet the requirements of the desired output geography. The first of these has been undertaken within the GAPS GIS, and the second using a series of custom programs (held on a standalone computer for security reasons).

18.4.1 Building blocks

Technically, it would be quite possible to construct a census database in which the geographical building block was the individual census return, suitably georeferenced. Such a database would have considerable potential for reaggregation of data to future output geographies, but it is neither ideal nor necessary for the creation of a new standard output geography. There are a number of reasons for this, of which the absence of appropriate geographical referencing is one of the most important, and the high cost of establishing such a system another. Although ADDRESS-POINT provides a 0.1 m resolution grid reference for each address in the Post Office's Postcode Address File (PAF), this will not be exactly the same as the list of households enumerated. The full extent and nature of the mismatch will not become apparent until processing of the 1997 Census Test areas is complete. Further, the single point reference provided by ADDRESS-POINT does not lead directly on to any particular mechanism for the geographical aggregation of data, as the topological relationships between the points are not known. A more appropriate building block would be the individual land parcel, but this would require definitive boundaries for all parcels, together with appropriate rules for dealing with all non-residential land parcels. The concept of such a definitive dataset has been considerably advanced by the discussions concerning British Standard 7666 (Cushnie, 1994), but these data do not yet exist in practice. It is interesting to note that the final definition of BLPUs used in BS7666 relates to the legal definition of property ownership, as this is the only formally-documented way of defining the boundaries of many land parcels (Pugh, 1992; Pearman, 1993).

Given the widespread demand for clearer integration between postal and census geographies, re-emphasized by Dugmore (1996), an obvious alternative building block is the unit postcode. This has the further advantage that postcodes were recorded on 1991 census returns. As long ago as 1987 the government committee of enquiry into the handling of geographical information recommended 'that the preferred bases for holding and/or releasing socioeconomic data should be addresses and unit postcodes' and urged that 'OPCS, in addition to the GRO(S), should ensure that the results of the 1991 Census of Population and any future censuses are available, subject to confidentiality, on a unit postcode basis' (p. 92) (DoE, 1987). In a review of 1991 Census geography, Clark and Thomas (1991) describe how the use of a postcode basis for 1991 geography was considered and rejected because of the absence of postcode mapping and, consequently, of the pictorial base which is used by enumerators to ensure full coverage. They concluded that the identification and mapping of the extent of each postcode would have been too time-consuming and expensive. The main obstacle to the use of unit postcode building blocks is still the absence of definitive boundaries, but this task can now be automated using newly available digital data.

Unit postcodes do not strictly cover the entire land surface, as they relate only to postal delivery points, but digitized boundaries exist for higher level units in the postcode hierarchy. ADDRESS-POINT locations must be used if a polygon-based unit postcode representation is to be devised which allocates all underlying addresses into the correct polygon. Generation of Thiessen polygons around individual ADDRESS-POINT locations and subsequent dissolution of boundaries between polygons having a common postcode is a useful approach, ensuring that every ADDRESS-POINT falls in the correct polygon (Martin and Higgs, 1997). In regularly laid out urban areas such an approach would in itself be sufficient to create reasonably sized and shaped unit postcode boundaries. A further constraint in the census output case is that unit postcodes must be split into smaller building blocks where they contain residents in more than one statutory area (i.e. ward level and above) in order to maintain precise statistics for statutory areas in the output data.

One of the key issues in polygon construction concerns the extent to which ancillary data should be used to control boundary placement. There are many existing digital datasets which might reasonably be incorporated into unit postcode boundaries, either as 'hard' features which unit postcode boundaries cannot cross, or as 'soft' features to which boundaries might be snapped if the feature fell between two groups of ADDRESS-POINTS having different postcodes. The problem of selecting appropriate features for boundary placement is most severe in remote rural areas, where there is no real relationship between postal geography and features existing on the ground. Existing postcode sector boundaries are clearly an example of a 'hard' boundary, but most other datasets, including 1991 ED boundaries, fall into the second category, and the problem then becomes one of prioritization. Integration of postcodes with existing boundary sets has the advantage that it becomes easier to describe the relationships between the different areal geographies, while alignment with physical features probably offers a more useful boundary configuration for ground-based activities such as catchment definition or market area analysis. An Arc/Info AML has been written which permits Thiessen polygons to be created around ADDRESS-POINT locations, clipped to the polygon boundaries of a controlling coverage, in this case the 1991 ED boundaries. Those property polygons sharing the same postcode can then be merged to create unit postcode polygons which incorporate ED boundaries where these comprise conveniently located features. ED boundaries incorporate many major topographic features such as main roads, rivers and railway lines, which also usually form boundaries between unit postcodes.

The single most comprehensive set of digital topographic features which may be used in an attempt to refine this procedure is Ordnance Survey's Land-Line product (Ordnance Survey, 1996), which is derived from three 'basic scales' of survey at 1:1 250 (urban), 1:2 500 (rural) and 1:10 000 (moorland). Coding of features is largely determined by their conventional cartographic representation on Ordnance Survey maps, and consists of such codes as 'water features', 'pecked road metalling' and 'general line detail'. Land-Line contains no explicit topology, and thus presents a very complex 'spaghetti' dataset (Figure 18.3 shows a portion of Petersfield in Hampshire), from which it is difficult to identify those features which should form part of 'natural' unit postcode boundaries (if such can be said to exist at all). The general line detail category in particular contains many diverse features, some of which might be useful in forming unit postcode polygons, and others which are effectively 'noise' (e.g. field boundaries and outbuildings). After a variety of experiments, an approach has been adopted here which demonstrates the

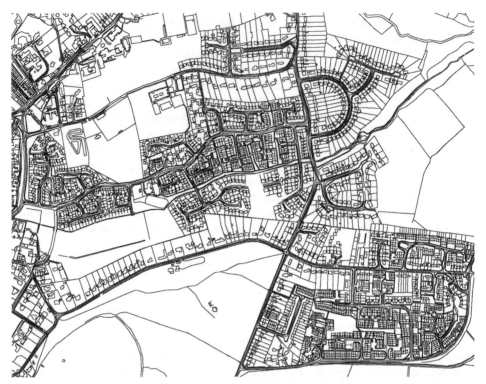

Figure 18.3 Subset of Land-Line information for Petersfield, Hampshire.

inclusion of selected Land-Line detail in artificially-generated unit postcode boundaries. The property-based Thiessen polygons, clipped to administrative boundaries, have been created as before, and form one input to the process. A small number of Land-Line features were selected for inclusion, comprising road centrelines, railway lines, water features (includes rivers) and general line features more than 25 m in length (to exclude outbuildings and other minor detail). These features were then merged with the property Thiessen coverage, providing a complex dataset containing all the features from which unit postcode boundaries might be drawn. Polygon topology was then created, and postcodes assigned into each polygon by intersection with the address layer. All boundaries between identical postcodes were dissolved to create a postcode boundary set which contained a number of (mainly very small) polygons to which no postcode was assigned. These were typically parts of Thiessen polygons cut off by irregular ground features, or areas of river or road divided by Thiessen boundaries, and they were eliminated by merging them with the postcoded polygon with which they shared the longest boundary. The output coverage was cleaned, snapping all near-coincident features, and any dangling nodes were removed. The resulting polygon coverage (illustrated by Figure 18.4 for the same area as Figure 18.3) is in most parts identical to that created by the simple clipped property Thiessen approach, but polygon boundaries which are close to real-world features such as road centrelines or river banks have been realigned along these. This procedure would be very costly in data and processing terms, yet the extraction of clean

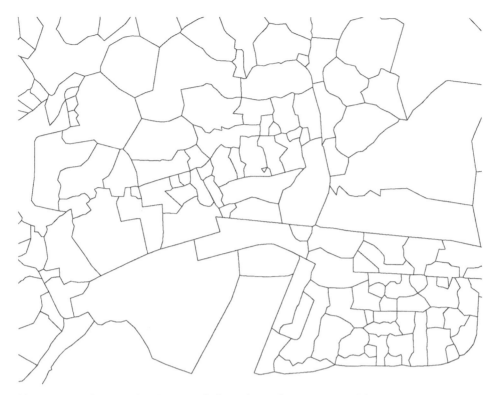

Figure 18.4 Generated unit postcode boundaries for area covered by Figure 18.3.

boundaries remains problematic, and is a generic problem in applications of this type. It is suggested that the simpler approach using constraining polygons such as 1991 EDs, and perhaps one or two other major features, is a more helpful approach to the creation of building blocks for a census output geography. The ED-constrained technique has been used to provide the building blocks for the work described in the following section of this paper. It is, of course, possible that commercial unit postcode boundaries may be available by the time of the 2001 Census which could be used directly as building blocks after intersection with statutory boundaries.

In general terms, there is a trade-off between boundary simplicity and inclusiveness. Automatic boundary generation approaches will tend to result in boundaries of varying point-density, frequently with minor irregularities which are artefacts of the constituent algorithms and ancillary data, and of no importance in the postcode geography. Unfortunately, all boundary generalization risks misallocation of some ADDRESS-POINTS, with the number of incorrect assignments increasing with the degree of generalization. This might be quite acceptable for thematic mapping purposes, but would be highly problematic for spatial analysis in which property locations were mapped back on to postcode polygons, such as epidemiological applications where individual records were to be related back to populations at risk. A further difficulty concerns non-contiguous postcodes. In rural areas particularly, some postcodes may contain properties at considerable distances from one another which are highly unlikely to be placed into adjacent polygons by any automated boundary-generation algorithm. To avoid the unacceptably irregular polygons which would be necessary to maintain strict contiguity, it must be accepted that some postcodes would require more than one polygon.

18.4.2 Output areas

Discussion is still under way concerning the most appropriate output geography(ies) for 2001, but custom output areas built from unit postcodes, as advocated here, are an attractive option not previously available (Rees and Martin, 1996). Having established a set of postcode-based building blocks, an automated procedure may be implemented for their combination into an 'optimal' output geography. A first task is the aggregation of population and other relevant census counts to the building block level. For this research, a special tabulation of selected unmodified 1991 SAS was undertaken by Census Customer Services at ONS. The SAS for most building blocks are below 1991 confidentiality thresholds, and could be differenced with 1991 EDs to create very small output units, and it has therefore been necessary to undertake output area construction on a standalone PC within ONS. In an operational context, such data could be fully integrated with the GIS, but isolated from the computer network. The building block contiguity matrix can be derived from the Arc/Info database by manipulation of the arc attribute tables, and this information has been transferred on diskette to the standalone PC.

In this research, a PC Fortran implementation of Openshaw's automatic zoning procedure (AZP) (originally outlined in Openshaw, 1977) has been undertaken, taking data from the GAPS GIS. Openshaw and Rao (1995) describe the re-engineering of 1991 geography using this approach with EDs as the building blocks, but this work concerns the explicit 'engineering' of a tentative 2001 geography. AZP is the simplest of the three zoning algorithms they investigate, but may offer the most effective solution in this context where many relatively small problems must be solved. Three aspects of output area design have been investigated: population size, geographical regularity and social homogeneity. In addition, a minimum population threshold of 100, and a requirement to prevent 'island' output areas have been implemented in these experiments. The actual modification and confidentiality procedures to be used for the 2001 Census are yet to be determined. This approach has been applied to seven wards drawn from 1997 Census Test areas in Birmingham, Hampshire and North Yorkshire. Sample results for the Sandwell area of Birmingham are presented in Martin (1997).

The AZP is an iterative algorithm which works by selecting an initial random aggregation of contiguous building blocks into OAs and evaluating potential building block swaps between adjacent OAs in order to optimize the selected objective function. When the best adaptation has been achieved, a new random starting configuration is chosen and the process begins again. It is therefore possible to affect the outcome by the selection and weighting of different objective criteria. In these experiments OA population size has been controlled by minimizing the squared difference between the target population and current population of each OA, boundary irregularity by minimizing squared boundary length, and social homogeneity by minimizing the squared difference between the proportion of households in the dominant class group and 1.0. The boundary length measure is an unsophisticated way of controlling shape, but was enforced by the necessity of separating the GIS and sub-publication SAS. Full integration would permit the use of better shape indicators by making boundary coordinates available to the zone design program. In this work, housing tenure class has been used as a proxy social homogeneity measure, although other measures could easily be developed. Population size and OA shape constraints could reasonably be implemented in most places in advance of the census, although social homogeneity would require processing of the building block counts before OA design.

Table 18.2 Summary results for selected 1997 Census test areas.

	No.	Min.	Max.	Mean.	St. Dev.	Class 1	Class 2
1991 ED	94	0	862	455	151	0.768	0.168
Postcode building block	1 302	0	279	35	33	0.857	0.093
Output area	177	106	574	249	79	0.790	0.162

A series of experiments have been undertaken in urban, suburban and rural test areas using different design criteria, and these are only very briefly reviewed here. Table 18.2 shows summary statistics for EDs, postcode building blocks and OAs across all test areas. In this example, all three objective criteria have been applied with equal weighting, and the best result from 500 iterations has been chosen. It can be seen that the building blocks offer enormous flexibility in OA construction, so that it is possible to meet the target population size of 250, while maintaining the minimum threshold of 100 (twice as high as for 1991 SAS), and increasing the average proportion of the dominant tenure class. The lower mean OA population results in many more OAs than 1991 EDs, which in itself increases homogeneity, and the smaller units would be particularly useful for the user in urban areas where ED sizes tend to be very large. There are no restricted output areas. The ability of this approach to achieve such improved characteristics so easily demonstrates the degree of sub-optimality of 1991-style EDs (which were, after all, designed as data collection units) for many output applications.

18.5 CONCLUSIONS

This paper has described the GIS used by ONS to establish an enumeration geography for the 1997 Census Test, and has addressed the design of a future census output geography. Output geography design has been described in two stages: the construction of building blocks and the creation of output areas. It has been suggested that the intersection of unit postcodes and statutory boundaries provide the most convenient set of building blocks, and a range of ancillary data has been used in attempts to generate boundaries for these units automatically. Land-Line offers the most detailed topographic information, but use of the data is complex, and the results unsatisfactory. In many areas simpler approaches, using merged Thiessen polygons around ADDRESS-POINT locations with constraining boundary sets, may offer acceptable solutions at much lower cost, and user requirements in this area should be further investigated. Existing algorithms offer practical means for the combination of these building blocks into OAs whose characteristics offer a number of improvements over 1991-style EDs, including much improved standardization of population size, shape and social composition, together with maximum integration with unit postcode geography and the absence of restricted OAs.

An entirely GIS-based approach to census geography design and management offers the potential to optimize collection and output geographies separately and to provide integrated digital products. Furthermore, the establishment of such a system as an on-going facility would permit the continual and immediate update of census counts for any spatial units resulting from local government, statutory or postal boundary changes in the future. It could readily be developed as the standard output from a system which main-tained a linkage between ADDRESS-POINT locations and individual census forms, pro-viding the basis for a future user-defined geographic enquiry system, as has already been suggested by (e.g.) Rhind *et al.* (1990). The establishment of such a geography design system is a reasonable and achievable goal for 2001, and is currently being considered in detail by ONS. There is a clear onus on the census-using GIS community to articulate its requirements now to ensure that 2001 census geography is designed both by, and for GIS.

Acknowledgement

This paper represents work undertaken during study leave from the University of South-ampton spent at the Office for National Statistics in Titchfield. The AZP program was developed after discussion with Professor Stan Openshaw. The assistance of ONS staff and use of their data and systems is gratefully acknowledged, but responsibility for all opinions expressed in the paper rests with the author.

References

BARR, R. (1993) Mapping and spatial analysis, in A. Dale and C. Marsh (Eds), *The 1991 Census user's guide*, 248–268, London: HMSO.

CHARLTON, M., RAO, L. and CARVER, S. (1995) GIS and the census, in S. Openshaw (Ed.), *Census Users' Handbook*, 133–266, Cambridge: GeoInformation International.

CLARK, A.M. and THOMAS, F.G. (1990) The geography of the 1991 Census, *Population Trends*, **60**, 9–15.

COOMBES, M. (1995) Dealing with census geography: principles, practices and possibilities, in S. Openshaw (Ed.), *Census Users' Handbook*, 111–132, Cambridge: GeoInformation International.

CRU/OPCS/GRO(s) (1980) *People in Britain – a census atlas*, London: HMSO.

CUSHNIE, J. (1994) A British standard is published, *Mapping Awareness*, **8** (5), 40–43.

DoE (1987) *Handling geographical information: the report of the Committee of Enquiry chaired by Lord Chorley*, London: HMSO.

DUGMORE, K. (1996) What do users want from the 2001 Census? in *Looking towards the 2001 Census*, 21–23, OPCS Occasional Paper 46, OPCS: London.

MARTIN, D. (1996) *Geographic information systems: socio-economic applications*, 2nd edn, London: Routledge.

MARTIN, D. (1997) From enumeration districts to output areas: experiments in the automated creation of a census output geography, *Population Trends*, **88**, 36–42.

MARTIN, D. and HIGGS, G. (1997) Population georeferencing in England and Wales: basic spatial units reconsidered, *Environment and Planning A*, **29**, 333–347.

MORPHET, C. (1993) The mapping of small-area census data – a consideration of the role of enumeration district boundaries, *Environment and Planning A*, **25**, 267–278.

ONS (1996) 1997 Census Test, *Census News*, **36**, 1–3.

OPENSHAW, S. (1977) A geographical solution to scale and aggregation problems in region-building, partitioning and spatial modelling, *Trans. Inst. British Geographers*, NS 2, 459–72.

OPENSHAW, S. (1995) The future of the census, in S. Openshaw (Ed.), *Census Users' Handbook*, 389–411, Cambridge: GeoInformation International.

OPENSHAW, S. and RAO, L. (1995) Algorithms for reengineering 1991 Census geography, *Environment and Planning A*, **27**, 425–46.

ORDNANCE SURVEY (1996) *Land-Line user guide*, Version 1.1, Southampton: Ordnance Survey.

PEARMAN, H. (1993) Designing a land and property gazetteer, *Mapping Awareness*, **7** (5), 19–21.

PUGH, D. (1992) The National Land and Property Gazetteer, *Mapping Awareness*, **6** (5), 32–45.

RAPER, J., RHIND, D. and SHEPHERD, J. (1992) *Postcodes: the new geography*, Harlow: Longman.

REES, P. (Ed.) (1997) *The debate about the geography of the 2001 Census: collected papers from 1995–6*, Working paper 97/1, Leeds: School of Geography, University of Leeds.

REES, P. and MARTIN, D. (1996) Flexible geographies and area aggregation: designing small areas for outputs from the 2001 census, in P. Rees (Ed.), *The debate about the geography of the 2001 Census: collected papers from 1995–6*, 11–29, Working paper 97/1, Leeds: School of Geography, University of Leeds.

RHIND, D.W., COLE, K.J., ARMSTRONG, M., CHOW, L. and OPENSHAW, S. (1990) *An online, secure, and infinitely flexible database system for the national population census*, SERRL Working Report 14, London: South East Regional Research Laboratory.

CHAPTER NINETEEN

Putting the past in its place:
the Great Britain historical GIS

IAN N. GREGORY AND HUMPHREY R. SOUTHALL

19.1 INTRODUCTION

Since 1994, a major project at Queen Mary and Westfield College has been constructing a historical Geographic Information System (GIS) covering the changing boundaries of major statistical reporting units of Great Britain from the early nineteenth century to the present. This is linked to a major database of census, demographic, economic, health, and electoral statistics, allowing mapping and analysis of the country's changing human geography from the beginnings of modern statistical data collection in the early nineteenth century to the present. This GIS is becoming a major resource in its own right, and it is hoped that it will be used to produce an electronic atlas, including animations and virtual-reality simulations, to mark the bicentenary of the first census in 2001 (Southall and White, 1997a). This paper discusses what the GIS contains, the technical challenges of building a GIS that includes a major time-variant element, and the analytical possibilities such a system creates.

19.2 WHY BUILD A HISTORICAL GIS?

From the early 1970s most centrally-collected statistics, such as the census, have been made available in digital form together with the necessary Digitised Boundary Data (DBD). Prior to 1971 few data were made available by the government in digital form, and almost all machine-readable datasets derive from *ad hoc* transcriptions made from the printed reports to meet the needs of a specific project. Given the abundance of machine-readable statistics and DBDs for the past quarter century, earlier periods tend to be neglected and the analysis of long-term trends has been severely hampered by the changing administrative geography. Our aim is to fill this gap by building a GIS containing both spatial and attribute data from the early 1970s back to the early nineteenth century. We define a historical GIS as a system containing the boundaries of areas as they changed over a significant period of time, stored in such a way that a user can extract spatial data for any part of the country and any date, and link this to relevant attribute data. We identify four reasons for building a system such as this: as a record of change, as an assembly of integrated information, as a powerful analytical tool, and as the basis for new media for visualizing changing geographies.

19.2.1   A record of change

Our GIS is a systematic record of administrative change containing a wide range of boundaries, each of which existed for some part of a long period, each tagged with the type of unit or units it bounded and the period over which it had legal status. This information is abstracted from a mixture of maps of administrative boundaries and textual records of boundary changes. This means that while one particular printed map provides a snapshot of boundaries on one particular day – say, 23 November 1909 – and another map shows the same units several years later, our research has located the precise dates of the changes that occurred in between the two maps, allowing an accurate coverage to be generated for *any* date. Our system is therefore a 'coverage generator' which enables a wide range of users to create the particular map they need, from local historians who simply want a choropleth map of a variable or variables in their local area, to geographers, economists, demographers who want to analyze long-run spatial trends. Dissemination of this material will be through the UKBorders census unit in Edinburgh.

19.2.2   An assembly of integrated information

Together with the spatial data we have assembled a very substantial major database of historical data which will cover the period from the earliest censuses to the 1960s; most of this material has been acquired through collaborations with other researchers. The database includes a wide variety of census data – vital registration data (births, marriages, and deaths), election results, statistics of economic distress (unemployment, the Poor Law, personal indebtedness) and so on. Our long-term aim is to assemble all recurring tabulations from the censuses of England and Wales and of Scotland, plus associated vital registration data, and we are collaborating with the 'Database of Irish Historical Statistics' project at the Queen's University, Belfast, who have already input a comprehensive assembly of Irish data and have a specialized Optical Character Recognition system with a proven ability to scan often-faded C19 printed sources. All our attribute data will eventually be made available through the Essex Data Archive.

19.2.3   A powerful analytical tool

While our system is of considerable interest to other researchers as a source of base maps and statistics, it also creates a vast analytical potential. Part of that potential is simply the mapping of previously unmappable datasets; work on the system began to enable the mapping of C19 Poor Law statistics, where a very large body of data for the individual Unions was little studied because their names often meant little, and much of the funding to extend the system has come via other researchers who wished to map specific datasets. However, the creation of a single large resource creates many new avenues for research, and we are exploring *three* specific directions: the construction of consistent time-series where reporting units change; the analysis of causal relationships where different variables are recorded for different systems of units; and the analysis of changing administrative structures.

First, basic demographic statistics, such as the age- and sex-structure of the population, numbers of births, and numbers of deaths by age group, have been systematically recorded since the mid-19th century and are little affected by changes in classification. However, the reporting units have changed drastically: until 1911, Registration Districts were the main reporting units; between 1911 and 1974 local government districts (LGDs) were used, meaning County and Municipal Boroughs, Rural and Urban Districts; more recently, very detailed Small Area Statistics are available for Wards or Enumeration Districts. Furthermore, these units were themselves subject to constant incremental changes, so that, for example, the Registration Districts of London were very extensively revised in 1901. This means that, while vital statistics have been gathered continuously since the introduction of civil registration in 1837, this 160-year period is rarely studied as a whole, and research on really long-run demographic change in Britain has often stopped in the 1840s, the principal focus being the ESRC Cambridge Group's work on parish registers (Wrigley and Schofield, 1981). The GIS contains comprehensive information on the changing units and by interpolating between the different areas for different dates it is possible to compute new series for a standard set of units, creating, for the first time, a set of consistent time-series at sub-county level. Obviously, other statistics such as occupational structure or cause of death are more problematic due to changes in classification. Once consistent demographic series have been constructed for the full period 1851 to 1991, a range of new indices relating to inter-censal demographic processes can be computed: we are using a cohort-survival method to estimate net migration rates for specific age- and sex-categories, which is showing that migration rates among young adults were both high and very sensitive to economic changes; and we are computing full life tables which will permit very detailed analyses of changing mortality.

Secondly, modern government statisticians endeavour to use certain standard sets of units for gathering a wide range of different information. In the past, a much wider range of units was employed and statistical analysis often appears impossible because explanatory variables are not available for the units of interest. A good example is research into electoral behaviour: voting statistics are available for parliamentary constituencies from 1832 onwards but little other information is available for these units; conversely, the mid-Victorian censuses used Registration Districts in most tabulations of socio-economic data. Political historians have dealt with this problem by massive aggregation, typically to county level (Miller, 1977; Wald, 1983), but the GIS can estimate socio-economic variables for constituencies, using forms of areal-interpolation (Goodchild et al., 1993; Flowerdew and Green, 1994).

Thirdly, the administrative structure of Britain has evolved massively over the past 150 years and the GIS documents this in great detail. Our boundary change databases include 450 changes to Registration Districts between 1840 and 1911, and 800 changes to sub-Districts, all taken from the Registrar General's Annual Reports and Decennial Supplements. We have nearly 5 000 changes to Local Government Districts from 1888 to 1973, taken from the Local Government Board's Annual Reports, and its successors. We are currently constructing a third database of changes to Civil Parishes, taken from the census printed reports from 1881 to 1971; this already contains 680 changes just for Wales, which contained only 1 102 of the 14 929 parishes in England and Wales in 1911. Our aim is to analyze systematically the geography and timing of these changes, examining the socio-economic and demographic pressures to which the system appears to have responded, and assessing how effectively the system did respond to massive long-run change.

19.2.4 New media for visualizing evolving geographies

While new statistical methodologies may advance academic knowledge, they influence popular understanding and political decision-making only rarely, slowly and indirectly. New methods for presenting our work, to students and the 'general reader', are therefore of equal importance. Another area of our work is investigating how best to develop an understanding of how long-run socio-economic change and demographic processes have influenced the geography of modern Britain. In this research we are exploring how best to adapt established tools for cartographic visualization to work with very long runs of historical data, and consequently how to replace static maps and visualizations on paper with animated graphics in a range of formats. While our long-term aim is to create a quite new atlas to appear in 2001, our current funding is enabling us to create an electronic version of an existing atlas, Langton and Morris' *Atlas of Industrialising Britain* (1986), drawing on both our GIS and the various statistics assembled by contributors to the original atlas. This will provide us with a full working prototype but should also create a resource of immediate usefulness: our electronic edition is designed to run within a standard Web browser, although it also requires various plug-ins such as Tcl/Tk, and we have permission to distribute copies to UK Higher Education Institutions at cost, via the CTI Centres for Geography and History (Southall and White, 1997a).

Our new time series for consistent units, discussed above, is forming the basis for animated cartograms which can represent both long-run changes in the distribution of population and the demographic processes driving them. We have modified a cartogram-generator developed by Daniel Dorling (Dorling, 1995) so that it interfaces directly with our attribute database, cycles through a whole sequence of dates, interpolating as required, and outputs a series of Postscript files which can then be assembled into a digital movie in a range of formats such as MPEG and AVI (Southall & White 1997b). The resultant images obviously cannot be included in a printed volume but can be viewed on our Web site:

http://www.geog.qmw.ac.uk/gbhgis

19.3 A BRIEF HISTORY OF LOCAL GOVERNMENT IN ENGLAND AND WALES

Between 1837 and 1911 the primary units used for the collection of demographic and socio-economic data at sub-county level were Poor Law Unions (PLUs) and Registration Districts (RDs). Although there were minor differences between these, they were broadly the same, being set up in the 1830s and falling into disuse in the early twentieth century. They were used for administering the poor law, collecting data on births, marriages and deaths, and as the main reporting units for the censuses from 1851 to 1911 (Hasluck, 1936). In the twentieth century the main administrative unit were local government districts: county and municipal boroughs, and urban and rural districts. These were formed under the Local Government Act of 1894 primarily by subdividing PLUs into urban and rural areas, so that while there were around 650 PLUs there were nearer 1 500 LGDs (Lipman, 1949). These became the main collecting districts for government, census, and other statistics until the 1972 Local Government Act was implemented. Most of these

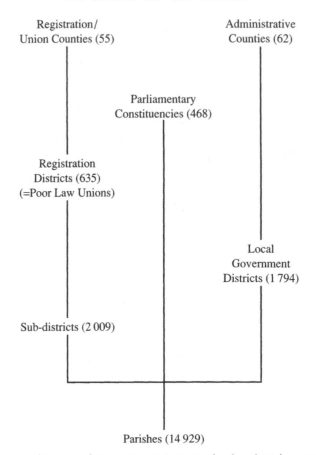

Figure 19.1 Principal Statistical Reporting Units in England and Wales, 1911.

areas have further subdivisions although the data available for them becomes more limited: RDs were subdivided into sub-districts of which there were about 2 000, while the fundamental building blocks for RDs and LGDs were parishes, of which there were approximately 15 000 in England and Wales. Parishes are further subdivided in many boroughs into wards. A simplified diagram of the structure of local government and related units in England and Wales, as they existed in 1911, is shown in Figure 19.1 although it should be noted that there were numerous exceptions and other complications.

Relatively little information for parishes or wards is available from official sources prior to modern Small Area Statistics. However, these units are important for three reasons. First, once we have captured the changing boundaries of civil parishes, a wide range of other boundary geographies can be constructed from them, by ourselves and by other users of our system. Secondly, our construction of new series for constant geographical units will be much more accurate if we can use parish-level population totals in our reallocation of data for higher-level units, rather than assuming that population had a constant density across these larger units. Thirdly, our collaborators include two major projects with parish-level data: the 1670s Hearth Tax project and a project at Essex University working with a *full* transcript of the 1881 census enumerator's returns, a database of *c*.30 million records including names, addresses and other attribute data for each individual resident in Great Britain in 1881.

A third higher-level geography which can be constructed from a parish-level GIS is parliamentary constituencies. These were changed not through a trickle of incremental changes but through wholesale re-districting at irregular intervals, which for most of our period reflected not only changes in the distribution of the population but major changes in who had the right to vote. Very detailed information on each revision is contained in the relevant Boundary Commission report, each of which contains a wealth of maps. That for 1832 eliminated the 'rotten boroughs', towns which by then had lost a former importance, concentrated in the south-west of England; that of 1866–8 followed the extension of voting to working-class adult males in the towns; that for 1885 marked a similar extension in rural constituencies; and so on. Unsurprisingly, each revision during the 19th century increased the proportion of constituencies in the north of England. Note that, while registration districts had been designed to combine towns with their hinterlands, for most of the period parliamentary constituencies were of two distinct types: county constituencies, each covering a substantial fraction of the rural population of a county; and borough constituencies, drawn tightly around a single town. The history of electoral geography has been neglected to an extraordinary degree, and we aim to change this through our mapping of constituencies, together with our methodology for synthesizing socio-economic statistics for these units.

19.4 BUILDING THE HISTORICAL GIS

Our system is a distinctively *historical* GIS: it is designed to produce accurate Arc/Info coverages for any date, precise down to the day, for any county or collection of counties in England and Wales, and to link this to any suitable attribute data set (census data, Poor Law statistics, etc.) held in our Oracle Relational Database Management System (RDBMS) for analysis and mapping. The system allows integration of coverages from different dates as the sources used have been standardized on a single series of source maps for a single date and projection. The final system, in its purest form, would only require four coverages plus software for data retrieval. These four coverages are: one holding PLUs and RDs, one holding LGDs, one holding wards and civil parishes, and one with parliamentary constituencies. The construction of this system has *four* elements:

■ *Gathering information* on the date and effect of boundary changes from maps and official reports.

■ *Digitizing a master map*, adding boundaries that had been abolished by changes and giving all features attributes to allow a coverage to be extracted for any appropriate date, area, and type of unit.

■ *Writing software* to allow easy data-retrieval based on a 'point-and-click' menu interface.

■ *Writing software and creating gazetteers* (look-up tables) to allow the map data to be linked to as wide a selection of attribute data as possible.

The largest part of our project has been capturing the changing boundaries. Our rather complex funding history has meant the project has developed in a number of stages, each adding an additional layer of complexity. This has meant some duplication of effort due to repeated passes through the same sources, but such duplication has been limited and starting with a relatively simple system enabled us to walk before we tried to run. Our work to date can be divided into three stages:

19.4.1   The Mark I GIS

Our initial funding covered a relatively simple GIS containing the *changing* boundaries of the *c*.630 PLUs and RDs between 1840 and 1910; this Mark I system was completed at the end of 1995 and has since been deposited with UKBorders. Although earlier researchers mapping RDs had relied on the poor quality maps in the census reports, after considerable investigation of the alternatives we decided that our main source for digitizing would come from the very end of our period: the second edition of the 2-miles-to-the-inch county administrative area maps published by the Ordnance Survey between 1907 and 1910. These show both PLUs and LGDs and, unlike earlier maps, used a single projection for the entire country. They do not include the National Grid, so we assemble the individual county sheets into a national coverage via a series of control points based on features near county boundaries, such as railway junctions and church towers, which consequently appear on more than one sheet. Additional maps from 1900 (2-miles-to-the-inch), 1888 (4-miles-to-the-inch) and 1851 (12.5-miles-to-the-inch sketch maps included in the printed census reports) were also used to help interpret changes.

Our main source of information on boundary changes were the *Annual Reports* and *Decennial Supplements* of the Registrar-General, which list all changes to RDs and sub-Districts from 1861 onwards with the precise date, area and population transferred; changes to PLUs were cross-checked from footnotes in the reports of the Poor Law Board and later the Local Government Board, while the limited differences between the two systems were listed in census reports. Pre-1861 sources are much more limited, but we drew extensively on a 'Statement of the names of the several Unions and Poor Law Parishes in England and Wales' (BPP [C.5191] 1887 LXX), which gave dates for the creation of each PLU and subsequent changes. Another cross-check was provided by the census reports from 1851 to 1911, each of which listed the population of each RD *as currently defined* for both the current and previous census, so that comparison of two census reports enabled us to identify all changed districts which had been redefined. For example, the 1911 census lists the 1901 population of Bridge RD in Kent as having been 10 971, where the 1901 census gives it as 12 384, so we conclude that a sub-area containing 1 413 people must have been transferred at some date between the two censuses; in this case we know from the Registrar General's reports that on 1 June the parishes of Thanington Within, Holy Cross and Westgate Without were transferred to Canterbury RD. In general, any change involving a transfer of at least 100 people was regarded as indicating a significant boundary change. In some cases we have been unable to locate any information on the actual boundary change, but these have been recorded and could be further investigated through local archives, etc.

We have also sought to include all boundary changes which covered a significant geographical area, clearly visible on the maps, even if fewer than 100 people moved; large changes in moorland areas could involve very few people. Some changes were easy to include in the GIS as they involved transfers of whole parishes, but smaller changes were harder to map. Where these occurred between 1888 and 1910, they could normally be identified by comparing the different series of maps, but before 1888 information becomes sparse, the census sketch maps being imprecise. In consequence, the further back you go from 1888 and especially from 1861 the reliability of the boundaries in the system declines. However, we have created a fully-documented record of all changes which were either unmappable or poorly mapped which will permit further investigation. The completed Mark I system embodies 416 distinct changes to the RD system.

The final challenge was how to enter boundaries and changes into a GIS in such a way that both storage and retrieval are efficient. This is made especially difficult by the fact that Arc/Info, like most other GIS software, embodies no concept of time nor change (Langran, 1992); unlike standard commercial RDBMSs, Arc/Info attribute tables lack even basic date fields. Our solution was to date-stamp every feature, both polygons (districts) and individual arcs (boundaries), with a start and end date (Vrana, 1990), and to store all features in one master coverage. Software was written in Arc Macro Language (AML) to extract a coverage for a given date and area. The minor differences between PLUs and RDs were also dealt with through arc and polygon attribute tables. Once the data were entered and attribute tables encoded, rigorous checking was essential to ensure topological consistency and accuracy at every date.

19.4.2   The Mark II GIS

As the Mark I system was nearing completion, new funding was obtained for an extended system which would convert our static coverage of 1911 Local Government Districts, captured while we were building the Mark I system, into a full record of changing LGDs between 1894 and 1973, and also create a system for parliamentary constituencies from 1832 to 1973. This became the Mark II GIS. The sources for working on LGDs are much better than for PLUs. Changes to boundaries were recorded in the *Annual Reports of the Local Government Board* and its successors, from which we have constructed a database of nearly 5 000 changes, while 2-miles-to-the-inch administrative area maps have been published at roughly ten-year intervals since 1910. Parliamentary constituencies are in many ways easier to handle, as the boundaries change far less often and in a far less *ad hoc* manner. Parliamentary constituency boundaries first appear on the 1920s series of administrative area maps, but the boundaries used often correspond to lines already captured. Details of earlier changes are taken from the Boundary Commissioners' Reports.

This Mark II GIS is essentially an enhancement of the Mark I system and is enabling us to experiment with analyzing and visualizing change over the full 1851–1991 period. It has been completed for the north of England, but further new funding is now enabling us to construct a completely new and far more detailed system.

19.4.3   The Mark III GIS

As we have already discussed, the system of *c*.15 000 Civil Parishes provided the most detailed administrative geography and the building blocks for higher level units. Although we are funded primarily to create base maps for projects concerned with the 1881 census and the 1670s Hearth Tax, we are striving to create something more generally useful, and in particular to create a full record of change between 1881 and 1973. By a fortunate coincidence, 1881 is the earliest date covered by comprehensive lists of parish boundary changes in the census reports, but the boundaries themselves are more problematic.

The first stage in construction of the Mark III GIS is a complete re-digitizing of the second edition of the 2-miles-to-the-inch administrative area maps, as used in the Mark I GIS; these are cartographically accurate, completely out of copyright and, unlike larger-scale maps, permit the digitizing of the entire country reasonably quickly. Mapping parishes means so many additional boundaries that re-digitizing is quicker than editing the Mark I system, and enables us to correct some problems with our earlier work. We previously input the main parish-level table from the 1911 census and, as each county is digitized, it is linked to that attribute table, checking that every parish which should appear is present. We also check that parish areas are within 5% of the acreages given by the census, except for coastal parishes where tidal areas complicate comparisons. The census table also gives the Registration District, sub-District and Local Government District to which each parish belonged, and can therefore be used as a look-up table defining higher-level geographies. This very complete reconstruction of the 1911 administrative geography is the bedrock on which our entire system is based.

There were three Acts of Parliament between 1871 and 1891 explicitly aimed at reforming the parish as an administrative units, resulting in huge numbers of boundary changes. The 1891 census report noted that 'of the 14 926 Civil Parishes of which the populations were given in the census returns of 1881, no fewer than 3 258 had their boundaries affected in the course of the next decennium' (Vol. IV, General Report, p. 2), primarily due to the Divided Parishes Act of 1882. In working backwards from 1911, we are therefore making extensive use of two sets of unpublished maps showing RDs, sub-Districts and parishes in the Registrar General's holdings at the Public Record Office (PRO class RG. 18), the first for 1891, the second being catalogued as covering 1871 but probably rather older; the latter set is also only 80% complete but can be supplemented for the north of England by boundaries marked on the printed First Edition 1-inch maps. For the remainder of the country we have located, at the Office of National Statistics and the Royal Geographical Society, two bound volumes of 1-inch maps marked up by hand to show parish boundaries, probably in the 1840s and drawing on tithe and enclosure maps. The maps in the PRO cannot be moved, so we plot out the digitized 1911 map on to tracing paper, enlarged to 1-inch scale; mark up differences at the PRO; and then re-digitize these differences from the tracing paper. The end result is a time-variant system containing all changes from 1881 to 1911, and some for earlier periods. There are some unavoidable problems where 1871 sheets are missing, or caused by our combining material from such a range of sources, dates and scales. However, our 1911 map provides a clear and very well-documented foundation and, as before, we have documented all known problems.

Mapping post-1911 changes is much more straightforward, using subsequent editions of the administrative area maps, although the sheer volume of changes to be incorporated is a problem. The fact that we are now dealing with 15 000 areas seriously complicates the process of extracting and handling the data, leading to some modification of our extraction software so as to access the area of interest more quickly.

19.4.4   ... And beyond

Mapping post-1973 boundary changes has a low priority as DBDs for the 1981 and 1991 censuses are already easily available, and the closer we come to the present the greater the copyright problems. We are extending the system to cover Scotland and Ireland through collaboration with UKBorders in Edinburgh and the Irish Historical Database project in Belfast. Our own long-term aim is to extend our parish-level system back from 1881 to 1801 by drawing on the sources mentioned already, the Ordnance Survey files in the PRO and the work of Richard Oliver and Roger Kain of Exeter University on the tithe and enclosure maps of individual parishes. However, it will not be possible to date changes with the same precision, and the main source of chronology will be the census reports themselves, as we link each parish-level table in turn into the GIS. The end-product will not only permit us to map all the censuses back to 1801, but will also provide a framework for research on much earlier periods, the system of parish units having been relatively stable over several centuries prior to the 19th century reforms.

19.5 LINKING ATTRIBUTE DATA

Once the GIS itself is ready, the next stage is to link the spatial data in Arc/Info with attribute data held externally in the Oracle DBMS. The Mark I system is tied together by names, the new Mark III system by ID numbers.

Linking in the Mark I system is done as follows: The names of PLUs and RDs are *mostly* unique, but combinations of name and county are always unique. However, using a relational join between Arc/Info and an external DBMS based on more than one field is difficult, so we add an extra field to the polygon attribute table (PAT) of the master coverage which contains a unique and standardized version of each district name. This is usually identical to the 'name' field, but in the remaining eight the name is followed by the county (e.g. 'WELLINGTON (SOMERSET)'). As the database contains many differ-ent tables taken from a wide variety of sources, spelling of place names varies consider-ably (for example St. Helen's, Saint Helens, etc.) and in order to preserve the integrity of the original source it is not considered desirable (or practical) to type in a standardized version at the data entry stage. To allow the names to be standardized to the GIS spelling we used a gazetteer, which contains every version of name and county found in the database and converts it to the standardized GIS name, allowing us to link an Arc/Info coverage to an Oracle table on one field without any 'holes' being caused by name problems.

Figure 19.2 (see colour section) was created using the Mark I GIS. It shows how the GIS permits new lines of analysis based on combining data from different sources, drawing on the census reports for 1901 and 1911 and the Registrar General's *Decennial Supplement* for 1901–10. The 1901 census tells us how many men aged 15 to 24 lived in each Registration District, and if none of them died or moved, and the District's boundaries did not change, exactly the same number of males aged 25 to 34 should have been living there in 1911. The *Decennial Supplement* tells us how many men in the age group died, while the GIS itself records boundary changes, so any remaining changes in the cohort size between 1901 and 1911 must result from in- or out-migration. The map shows clearly the very significant percentages of this particularly mobile age group leaving many rural areas, and moving into the developing coalfields, the London suburbs and various resort areas. One reason this calculation is of particular interest to us is that the necessary data are potentially available for every inter-censal period between 1851 and the present.

The parish-level system raises extra problems, owing to its size. Producing a unique and meaningful name for every civil parish that has ever existed is impractical, and we are therefore using ID numbers. These are six-digit numbers which for parishes in exist-ence in 1911 indicate their position in the parish-level census table: the first three digits indicate the RD, the second the parish within the RD. The ID has to be added manually to each label point during digitizing, but thereafter we can access names and other attribute data held in Oracle. The 1911 census table therefore forms the basis of an initial parish-name gazetteer, to which we can add variant spellings, etc. Parishes which dis-appeared earlier or appeared later are given IDs from outside the 1911 sequence and their names added to the gazetteer. Figure 19.3 (see colour section) shows an early result of this work, mapping one of the more interesting variables in the 1911 parish-level table to show the remarkable growth of the South Wales coalfield. The absence of 'empty' polygons shows our success in linking data to the GIS.

19.6 CONCLUSIONS

The Great Britain Historical GIS had quite modest beginnings and has developed through a number of stages. The first version was designed simply to solve the problems of an individual researcher, although the need to map bi-annual poor law statistics rather than ten-yearly census data meant that the problem of continuously changing boundaries had to be addressed directly, laying the foundations for a much more general solution. The much larger parish-level system has been funded mainly to meet the requirements of a number of individual research projects with a primarily demographic focus, but here again what we are building is proving of much wider value:

- To historians of particular localities, accessing new on-line resources providing base maps and attribute data.

- To the Office of National Statistics and, we hope, to many schools and colleges, as the basis for an innovative atlas marking 200 years of the census.

- To the Royal Commission on Historic Monuments for England, as a framework for contextualizing historic industrial sites.

- To the Public Record Office, we hope, as the basis for directing genealogical researchers to files for the relevant Registration District, based on known placenames.

- To archivists and perhaps even art historians, as the basis for placename authority lists and thesauri of geographical names.

- And not least, to ourselves, by making possible quite new analyses of truly long-term social, economic and demographic changes as they affected the intricate geographical fabric of the British Isles.

Acknowledgements

The Great Britain Historical GIS has been funded (so far!) by grants from the Economic and Social Research Council, Joint Information Systems Committee, Population Investigation Committee, the British Academy, the Nuffield Foundation and the Aurelius, Leverhulme, Pilgrim and Wellcome Trusts. Our visualization work has made use of code supplied by Daniel Dorling (Bristol) and Jason Dykes (Leicester) while our attribute database includes datasets supplied by collaborators too numerous to list here.

The digital boundaries used in the figures are Crown and ED-LINE Copyright, and were provided as part of the ESRC/JISC Census of Population Initiative.

References

DORLING, D. (1995) *Area cartograms: their use and creation*, Concepts and Techniques in Modern Geography series, Norwich: Environmental Publications, University of East Anglia.
FLOWERDEW, R. and GREEN, M. (1994) Areal interpolation and types of data, in S. Fotheringham and P. Rogerson (Eds), *Spatial Analysis and GIS*, London: Taylor & Francis.

GOODCHILD, M., ANSELIN, L. and DEICHMANN, U. (1993) A framework for the areal interpolation of socio-economic data, *Environment & Planning A*, **25**, 383–397.

HASLUCK, E. (1936) *Local Government in England*, Cambridge: Cambridge University Press.

LANGRAN, G. (1992) *Time in Geographic Information Systems*, London: Taylor & Francis.

LANGTON, J. and MORRIS, R. (1986) *Atlas of Industrialising Britain 1780–1914*, London: Methuen.

LIPMAN, V. (1949) *Local Government Areas 1834–1945*, Oxford: Basil Blackwell.

MILLER, W.L. (1977) *Electoral dynamics in Britain since 1918*, London: Macmillan.

SOUTHALL, H.R. and WHITE, B. (1997a) Creating an Electronic Historical Atlas of Britain, *Geocal*, **16**, 3–6, Leicester: CTI Centre for Geography.

SOUTHALL, H.R. and WHITE, B. (1997b) Visualising Past Geographies: the use of animated cartograms to represent long-run demographic change in Britain, in A. Mumford (Ed.), 95–102, *Graphics, Visualisation in the Social Sciences*, Loughborough: Advisory Group on Computer Graphics.

VRANA, R. (1990) Historical Data as an explicit component of land information systems, in D. Peuquet and D. Marble (Eds), *Introductory Readings in Geographic Information Systems*, 286–302, London: Taylor & Francis.

WALD, K.D. (1983) *Crosses on the ballot: patterns of British voter alignment since 1885*, Princeton: Princeton University Press.

WRIGLEY, E.A. and SCHOFIELD, R.S. (1981) *The population history of England 1541–1871: a reconstruction*, London: Edward Arnold.

CHAPTER TWENTY

Remote sensing and regional density gradients

VICTOR MESEV AND PAUL LONGLEY

20.1 INTRODUCTION

During the last 20 years urban studies have eschewed the task of creating generalizations capable of linking the morphologies of urban settlements to measures of their socio-economic functioning. Disenchantment with the metaphor of cities as spatial mosaics, and with classic location theory, has in large part been attributable to difficulties in measuring and manipulating geographical objects, problems in linking spatial form to social process, and constraints upon integration of diverse data sets with different data structures (Masser and Blakemore, 1991). These conditions are, however, rapidly changing. Continued and cumulative developments in geographic information-handling technologies and the emergence of rich sources of digital data unavailable hitherto have taken place during the last decade. The changes in the morphology of cities and of settlement distributions have been equally dramatic during this same period, and there is now a sense that we are entering a period of urban change more profound than any since the innovation of the industrial city (Castells, 1996). Tangible changes in built form caused by the demise of strict retail hierarchies, new movement patterns caused by changes in the labour market, and increased household fission giving rise to new and different demands for housing are all contributing to urban change. This is making it ever more essential that we understand the detailed interactions between built form and human activity, and their effects upon the structuring of cities and city systems.

In this chapter we attempt a preliminary investigation of the relationship between the built form and socio-economic functioning of the urban settlement hierarchy focused upon Norwich. To achieve this, we will use GIS-based data models derived from two different representations of the UK Census and upon Landsat TM imagery. We seek also to use the results of this analysis to evaluate established thinking about urban density gradients, extrapolated to a regional settlement system (Parr, 1985a; 1985b).

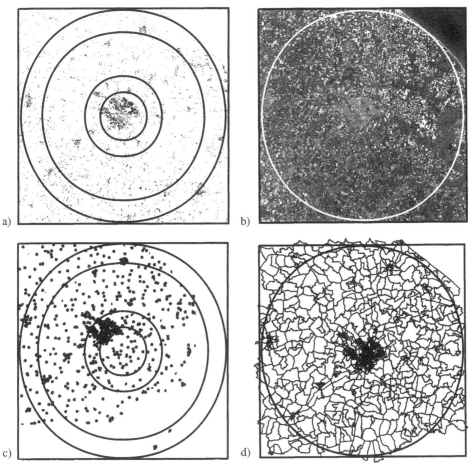

Figure 20.1   Multiple representations of the Norwich region: a) residential land use classification; b) Landsat-5 TM image; c) point-based census surface; d) enumeration districts.

20.2 CONVERSION

Information on settlement structure can be represented in a variety of formats, both spatial and aspatial. Regional density gradients have historically been measured from population censuses represented by zonal tracts (Figure 20.1d). The Census simply measures the number of residential properties within each tract, but this bears no relation to the underlying physical morphology. Remotely-sensed data (Figure 20.1b), however, can be used to interpret the land-cover structure of residential buildings (Figure 20.1a), yet cannot easily determine socio-economic characteristics (Donnay and Barnsley, 1997). What is needed is a methodology that links the fine spatial resolution, temporal flexibility, and the ability to locate residential buildings precisely from remotely-sensed imagery with aspatial information from the Census. Figure 20.2 is a diagrammatic representation of the flow of analysis needed to convert both census and image data, linked by dasymetric analysis, into density profiles. As a check on census and image data, an intermediary in the form of a point-based interpolated census surface (Figure 20.1c) is also converted into a density profile.

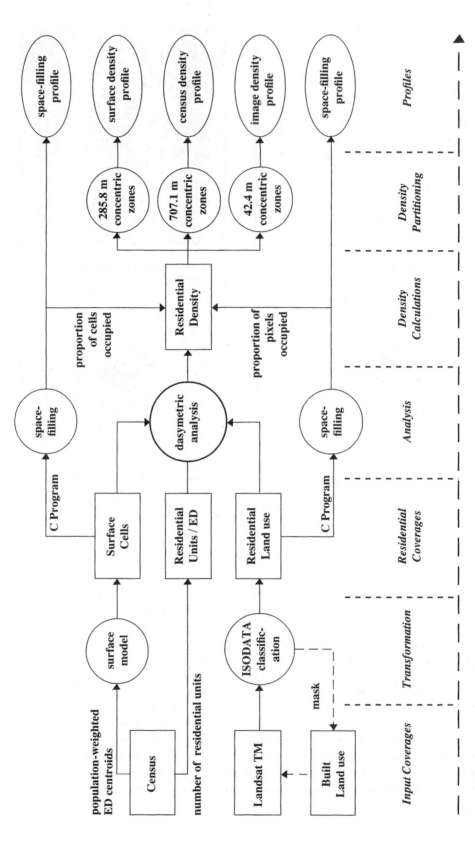

Figure 20.2 Flow of operations needed to convert each data set into space-filling and density profiles.

The procedure begins with the extraction of information on all residential units within the spherical limits of the Norwich study region. From the Census residential units can be accessed through the interactive, menu-based SASPAC system at the University of Manchester. The number of residential units per census tract, or ED (enumeration district), is extracted along with the ED's population-weighted centroid (located at the densest point of the largest built-up area). Conversely, from the Landsat-5 TM image (taken on 15/7/89) (Figure 20.1b), residential land use is extracted by straightforward, spectra-only, unsupervised ISODATA ERDAS classifications. The classifications are conducted under a two-stage hierarchical stratification. The first classifies built land cover, which is then used to mask out all non-built areas from the original image; the second classifies residential from non-residential land use, but only from within the confines of the masked built areas of the image. This repeat classification allows more classes to be generated (and hence greater contrast between residential and non-residential land use) from built pixels that have very similar spectral values in most bands. Although supervised urban image classifications have been documented to produce higher accuracies, including ones aided by ancillary information (Mesev *et al.*, 1995), the scope of this work-in-progress is to identify the possible role of remote sensing in determining regional density gradients rather than advancements in image classification. Therefore, unsupervised classifications are seen here as sufficient to generate rapid small-scale coverages of residential land use from large multispectral images.

Once residential data from both the Census and Landsat image are extracted, two space-filling and three density profiles are produced. Two of the density profiles are derived from census data: the first is generated using conventional zonal (ED) data and will be called the 'census density profile'; and the second is determined through the use of a surface model and will be called the 'surface density profile'. The third 'image density profile' is derived from the Landsat TM image.

The surface model is a point-based, areal, interpolation algorithm which uses a distance-decay function between ED centroids in order to generate a regular grid of cells. The algorithm was devised by Bracken and Martin (1989) in order to produce a finer and more realistic model of residential geography which was independent of zonal census tracts, but which did not require detailed information on the physical morphological structure of settlements. It is straightforward to calculate the 'crude' residential density at the ED level, for input into the calculation of the 'census density profile', by dividing the number of residential units by ED area. Both surface cells and classified image pixels are two-dimensional binary structures that record the presence (coded 1) or absence (coded 0) of residential development. (The surface model, in fact, generates continuous per-pixel density estimates, but these have been reclassified as nominal estimates for the purposes of this analysis.) Dasymetric mapping is used to link the number of residential units from the Census to the location of each surface model cell or pixel (see Langford *et al.*, 1991). The process first involves determining the proportion of an ED a cell or pixel occupies, and then allocates to that cell or pixel the corresponding proportion of the number of residential units from the ED.

Dasymetric analysis is again used to recalculate density values for all profiles into a series of regular concentric zones radiating out from Norwich Castle (which is taken as the historic centre of the city's development). The width of the concentric zones that are used to create the profiles is equal to the diagonal lengths of cells (200 m), pixels (30 m) or, in the case of the census profile, an imposed 500 m interval based on the average area of an ED. Converting regular grids into spherical zones preserves the equi-distance assumption that is critical in calculating density profiles. In order to make more direct comparisons, the surface and image density profiles are also represented by the coarser 500 m zone intervals.

Regional density gradients are converted from spatial data in Figure 20.1 using the methodology in Figure 20.2 and the following calculations. For each concentric zone i (where $i = 1, 2, 3, \ldots, n$) density d is measured incrementally by finding the ratio between the number of residential units H_i and the zone's area of occupation $A_{i,c}$,

$$d_i = \frac{H_i}{A_{i,c}} \qquad (20.1)$$

This is expressed as the number of residential units per kilometre squared. The area of each concentric zone, or ring, is calculated by determining the radius r of i and subtracting the area of the previous zone. For each zone to represent only occupied areas, the total number N of unoccupied cells or pixels $x_{i,o}$ are also subtracted, so that,

$$A_{i,c} = \pi(r_i^2 - r_{i-1}^2) - (Nx_{i,o}^2). \qquad (20.2)$$

Note that as the Census is a continuous coverage, its occupied areal resolution is undefined and therefore $A_{i,c} = \pi(r_i^2 - r_{i-1}^2)$. The numerator of (1) is calculated by dasymetrically linking each $x_{i,o}$ with spatially-corresponding residential information from census EDs, so that H_i represents the proportion of residential units from each ED that fall within occupied cells or pixels, $Nx_{i,c}$ within each zone.

Figure 20.2 also illustrates how two further profiles are generated. These are derived directly from the space-filling calculations for both the surface and image residential data, and represent the rate of urban space-filling in a strictly two-dimensional perspective. In (1) and (2) they are used to determine $Nx_{i,c}$ and $Nx_{i,o}$.

20.3 DISCUSSION

Settlement-density gradients have traditionally been used in a wide variety of applications to examine processes such as utility maximization (Muth, 1969), suburbanization (Mills, 1992), and decentralization of functions (Anas and Kim, 1992); as well as extended to the region (Parr, 1985a; 1985b). Moreover, the mathematical function which is used to generate density gradients has also been the focus of much deliberation and testing (Zielinski, 1980; Mills and Tan, 1980). However, although applications and statistical mechanics are no doubt significant, there has been a distinct lack of importance attached to the type and quality of spatial data from which gradients are generated and processes interpreted. We will now discuss how choices in the spatial entity, as well as the type of function, directly affect the form of regional density gradients.

20.3.1 Measurement (gross vs net density)

The 'census density profile', like many others used previously, is derived from population census tracts. Each tract is of variable size and contains census information in an aggregated form. These tracts constrain the conventional data model of residential densities to a mosaic, each tile of which is characterized by uniform within-area density. Because residential units are distributed uniformly within the tract at equal density, there is no intrinsic relation between density and the structural morphology of a settlement. We term all such measurements *gross* densities of residential units.

Table 20.1 Data parameters for the three data sets.

Parameter	Census	Surface model	Satellite image
Data Type	zonal	raster	raster
Spatial Resolution (x)	variable	202 m	30 m
Number of Concentric Zones (N_i)	54	95	633
Zone dimension (i) $(i = 1,2,3, \ldots ,n)$	707.1 m	285.7 m	42.4 m
Maximum Radius (R_m)	27.0 km	27.01 km	26.89 km
Total Area (A) πR_m^2	2 290.22 km^2	2 291.92 km^2	2 271.60 km^2
Number of Occupied Cells (Nx_c)	–	7 689 cells	78 881 pixels
Area Occupied (A_c)	2 290.22 km^2	313.74 km^2	70.99 km^2
% Area Occupied	100.00	13.69	3.13
Number of Residential Units (H)	176 147	175 686	175 703
Density of Residential Units (d) H/A_c	76.91	599.97	2 475.04

On the other hand, the surface and image models each filter out unoccupied space (i.e. non-residential land use) and hence replicate the physical morphology of urban areas. By overlaying the conventional data model of the census (with its known ED distribution of residential units) on to each of these models, it is possible to create two further estimates of the density of residential units by assuming that all occupied space within each ED contains a uniform density of residential units[1]. When residential numbers are combined with residential land use this produces *net* densities. The surface model distributes residential units from the population-weighted centroid of one tract towards population-weighted centroids of neighbouring tracts. As centroids are deliberately located within residential areas, the resulting surface cells determine an approximate distribution of the residential geography of a settlement. However, cells are still based on census data and have no comprehensive relation with the underlying structural morphology. Cadastral and property data sources aside, this can only be determined by data that are primarily designed to measure physical land cover, i.e. buildings, vegetation, etc. Information from remotely-sensed data is predominately determined by the reflectivity of energy from physical objects on the Earth's surface. Image classification is the grouping of spectral values that represent the characteristic mix of land cover associated with residential land use.

The differences between gross and net density are further highlighted in Table 20.1. Note as the census represents total occupied space, this is over 7 times more than the Census, and over 32 times more than the image. This means that, while all are ultimately constrained to known census totals, the smaller areas of likely occupied space for the surface and image means that density values are much higher than the census profile. Between the two datasets measuring net density, only the image can be reliably used to determine the physical structure of residential buildings and the amount of space these buildings occupy. Figure 20.3 illustrates the extent to which the the image and surface model differ in the amount of occupied space, and confirms that, when measuring residential land cover, models based on the Census are no substitute for models based on land properties (image). Its also worth noting that the higher spatial resolution of the image has meant that 633 concentric zones were generated, over 6 and 11 times as many as the surface and census, respectively. More zones allows greater variations in density to be measured, and this is illustrated in the more irregular form of the profiles.

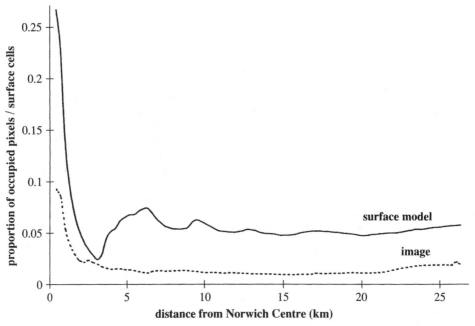

Figure 20.3 Space-filling profile of Norwich region using image and census surface data at spatial resolution of 283 m.

20.3.2 Profiles (goodness-of-fit)

Work on density gradients based on census tracts have traditionally favoured the use of the negative exponential function (Clark, 1951). This function assumes that population density at distance r from the centre of the settlement (where $r = 0$) declines monotonically according to the following relationship,

$$d_r = K \exp(-\lambda r) \tag{20.3}$$

where K is a constant of proportionality which is equal to the central density d_0 and λ is a rate at which the effect of distance attenuates: so that if λ is large, density falls off rapidly; if it is small, density falls off slowly. By the late 1970s, exponential density functions had been estimated for many metropolitan areas in many countries. The appeal of the function lay in its analytical expediency in problems of economic optimization, as well as in its simple linear representation of density attenuation more discussion in Zielinski (1980). However, over more recent years, processes such as residential sub-urbanization and the decentralization of industry and services in particular, have produced density gradients that are far from simple or linear. Furthermore, regional gradients are characterized by a central large settlement that invariably produces very high densities at this core area. What is needed is a function that takes both of these into account by being more flexible at both the periphery and core. In this paper, we argue that the inverse power function is more appropriate than the negative exponential for representing more profiles in contemporary urban structures. Instead of a constant rate of decline with distance, the inverse power function predicts higher densities at the core (central settlement) and a less steep decline at the periphery (suburbanization and decentralization). The function is expressed as,

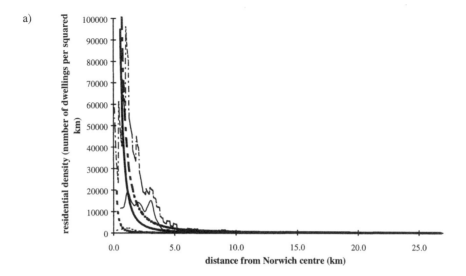

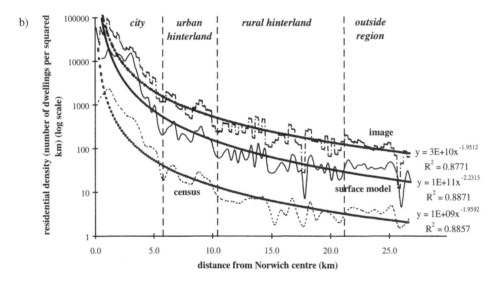

Figure 20.4 Residential density profiles at finest spatial resolutions available: (a) absolute; (b) log scale.

$$d_r = K\, r^{-\alpha} \qquad (20.4)$$

where K is again the constant of proportionality as in (3) but not defined where $r = 0$, and α is the parameter on distance.

The results of applying the inverse power function to the three types of data are shown in Figures 20.4 and 20.5. Figure 20.4 shows the three profiles plotted at the finest resolution available, and Figure 20.5 at standardized concentric zones at the coarsest resolution (the census profile at 500 m). Both figures demonstrate the expected sharp decline of residential density with distance from the central main settlement. However, density

a)

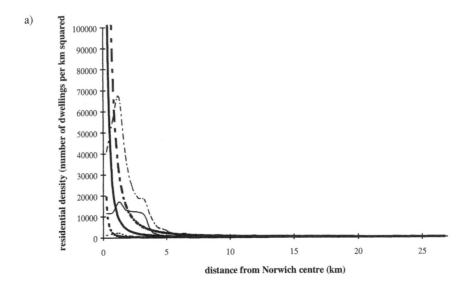

b)

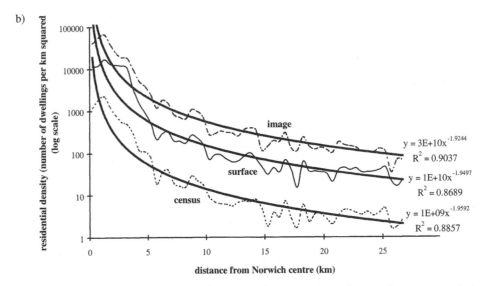

$y = 3E+10x^{-1.9244}$
$R^2 = 0.9037$

$y = 1E+10x^{-1.9497}$
$R^2 = 0.8689$

$y = 1E+09x^{-1.9592}$
$R^2 = 0.8857$

Figure 20.5 Residential density profiles at common spatial resolution of 500 m: (a) absolute; (b) log scale.

values and the rate of decline vary considerably, and are very much the result of measurement and resolution differences as we have already seen. Because both the image and surface profiles are measures of net density, they are at a much higher level of resolution than the census profile. If fixed numbers of residential units are constrained to known likely areas of occupied space, this leads to higher density estimates across all of the concentric zones. In comparing the two net density profiles, that derived from the image indicates higher density simply because its finer resolution (30 m compared with 200 m for the surface) results in still smaller occupied. Remember the residential numbers are the same for all three profiles. Apart from higher generalization in the image and surface profiles, the shape and order of the gradients between those in Figure 20.4 and those in Figure 20.5 do not vary considerably. This shows that, despite standardizing for sampling intervals, measurement and resolution differences are transferable across different scales.

Looking at Figures 20.4a and 20.5a it should be clear that all three density profiles are far from linear. They are all heavily skewed towards the Norwich settlement, as it would be expected in a region dominated by a single large settlement. The image profile is so skewed that the use of absolute density on the y-axis means that values beyond 5 km are completely suppressed. However, if density is plotted on a log scale, variations in the entire profile can then be seen (Figures 20.4b and 20.5b). Generally, these profiles are in a parallel sequence with virtually no interceptions between them. The order, as expected, is a function of resolution, or the proportion of occupation. However, there seems to be a striking similarity in the shape and pattern between the image and surface profiles. Remember, these profiles were generated completely independently of each other, with the only link being a dasymetric process in order to bestow the same residential numbers from the Census to either pixel or cell (in other words, a constant). This similarity is interesting in that it seems to suggest that links are broachable between land cover (image) and land use (census) data.

The degree of non-linearity in each of the three profiles can be more formally assessed by goodness-of-fit trend lines. Figure 20.4b illustrates how inverse power functions (variable base and constant negative exponential) with r^2 values of over 0.87 are reasonably good representations of the negatively-skewed density profiles. The only problem is that power functions become asymptotic towards 0 and are therefore poor indicators under 1 km from the core (Batty and Kim, 1992). Nonetheless, once these functions are fitted, it is then interesting to see whether deviations are evident – in other words, differences between the power functions and the empirically-derived profiles. The largest deviations are undoubtedly between 1 km and 5 km from the Norwich centre. This is where density values are much higher than those predicted by the power functions. The range also coincides with the full extent of the city of Norwich, whose size is far larger than the rest of the region. Future research may well compare this deviation with others from similar single-city regions to determine whether there is a trend and whether the degree of deviation is a reflection of the size of the sphere of influence – or is a common artefact of the measurement methods. Conversely, a large area of the region that has lower density values than predicted lies between 10 km and 21 km from the centre. From Figure 20.1 this is a predominantly rural hinterland with very few service settlements. In between the city and rural hinterland lies an area between 5 km and 10 km from the centre. It is characterized by density values that are closely predicted by the functions, and represents the transitional stage between the city and surrounding rural hinterland. Beyond around 21 km density values are again above the power functions. Although still rural, this area now includes a few more large settlements with higher-order services that can compete with Norwich. These distinctive regions are also marked areally on Figures 20.1a and 20.1c and confirm the changes in residential land cover.

20.4 CONCLUSIONS

This work has demonstrated how remotely-sensed data can be used as a new source in the reappraisal of established notions of regional density gradients. It has revealed how spatial, and not simply spectral, information from satellite imagery can represent the degree of space-filling and spatial irregularity in residential development, as well as examine the level and rate of attenuation from the urban core.

Classified satellite data provide an invaluable means of detecting urban land use, which can become even more profitable when used as the basis for intensive modelling. Density profiles can be generated routinely by spatially extensive, multi-temporal, digital remotely-sensed data. There is no other data set which has such properties of generality and allows the production of consistent land use classifications for various types of settlements to be explored through time. Many other spatial data sets are based on census information and, therefore, can only be effectively updated every ten years. Furthermore, classified satellite imagery is also able to illustrate distinctive and detailed characteristics in the profiles of urban areas. Many of these characteristics are as much a result of the fine resolution of the satellite data as a symptom of local constraints in urban development arising from factors such as physical barriers and restrictive planning policies. Nevertheless, the general pattern in these profiles seems to suggest that density is much higher than expected close to the core, as predicted in the urban hinterland, lower in the surrounding extensive rural hinterland, and higher at the periphery where other large settlements may come into effect.

There are many policy implications, particularly those involving accessibility and energy use as well as questions of residential and economic spatial shifts. Rapid methods of urban classification and measurement will help towards the understanding of how cities fill their available geographical space, and this in turn could inform more efficient planning and infrastructure provision policies.

Finally, the discussion of urban density over the last 50 years has been flawed because researchers have not had good data at the micro level, have not been able to relate density to built form, and have not been able to manipulate the geographic system to find the correct scale at which density should be interpreted. We are now in a position to re-appraise the relevance of density, and we believe that the powerful techniques at our disposal are able to make new sense of the mosaic of densities that comprise contemporary cities and regional settlement distributions. It is our belief that there are some very strong patterns to be uncovered, and that these patterns are simply obscured in most analysis which is scientifically loose in its control of the spatial system and its attributes. Examining densities using administrative geography is simply no longer good enough. We are not suggesting that census-based data should not be used at all. Instead we posit the role of digital, in particular remotely-sensed, data within integrated methods for determining urban models such as density gradients. Part of our wider quest of this research is to demonstrate the need for better scientific standards in responding to such issues.

Acknowledgement

Work for this paper was funded by Research Fellowship number: H53627501295 from the UK Economic and Social Research Council (ESRC).

Note

1   In the case of the surface model, this entails discarding of information pertaining to *continuous* residential densities: however, this was deemed unlikely to affect significantly the results of the exploratory analysis conducted in this paper.

References

ANAS, A. and KIM, I. (1992) Income distribution and the residential density gradient, *J. Urban Economics*, **31**, 164–180.

BATTY, M. and KIM, K.S. (1992) Form follows function: reformulating urban population density functions, *Urban Studies*, **29**, 1043–1070.

BRACKEN, I. and MARTIN, D.J. (1989) The generation of spatial population distributions from census centroid data, *Environment and Planning A*, **21**, 537–543.

CASTELLS, M. (1996) *The Rise of the Network Society*, Oxford: Blackwell.

CLARK, C. (1951) Urban population densities, *J. Royal Statistical Society (Series A)*, **114**, 490–496.

DONNAY, J-P. and BARNSLEY, M.J. (1997) *Remote Sensing and Urban Analysis*, London: Taylor & Francis (forthcoming).

LANGFORD, M., MAGUIRE, D.J. and UNWIN, D.J. (1991) The areal interpolation problem: estimating population using remote sensing in a GIS framework, in I. Masser and R. Blakemore (Eds), *Handling Geographical Information: Methodology and Potential Applications*, 55–77.

MASSER, I. and BLAKEMORE, M. (Eds) (1991) *Handling Geographical Information: Methodology and Potential Applications*.

MESEV, T.V., BATTY, M., LONGLEY, P.A. and XIE, Y. (1995) Morphology from imagery: detecting and measuring the density of urban land use, *Environment and Planning A*, **27**, 759–780.

MILLS, E.S. (1992) The measurement and determinants of suburbanisation, *J. Urban Economics*, **32**, 377–387.

MILLS, E.S. and TAN, J.P. (1980) A comparison of urban population density functions in developed and developing countries, *Urban Studies*, **17**, 313–321.

MUTH, R. (1969) *Cities and Housing: The Spatial Pattern of Urban Residential Land Use*, Chicago: University of Chicago Press.

PARR, J.B. (1985a) A population-density approach to regional spatial structure, *Urban Studies*, **22**, 289–303.

PARR, J.B. (1985b) The form of the regional density function, *Regional Studies*, **19**, 535–546.

ZIELINSKI, K. (1980) The modelling of urban population density: a survey, *Environment and Planning A*, **12**, 135–154.

CHAPTER TWENTY-ONE

Modelling change in the lowland heathlands of Dorset, England

ABIGAIL M. NOLAN, PETER M. ATKINSON AND JAMES M. BULLOCK

21.1 INTRODUCTION

This chapter presents the results of an object-based analysis of the spatial dynamics of heathlands in Dorset, England undertaken using Geographical Information System (GIS). The long-term aim of the research is to examine the *ecological* dynamics of heathland at the individual patch level to produce a statistical predictive model of heathland dynamics. This model should provide a simple set of guidelines for use by heathland managers. This chapter outlines the first stage in producing this model, involving both cell-based and patch-based approaches to examining heathland change.

The lowland heathlands of Dorset were created in the Mesolithic period when humans felled the forest and allowed animals to graze, so preventing the re-establishment of the climax vegetation. They provide a habitat for many important species such as the Dartford warbler (*Sylvia undata*), the sand lizard (*Lacerta agilis*), the smooth snake (*Coronella austriiaca*) and the marsh gentian (*Gentiana pneumonanthe*) (Webb and Haskins, 1980). Furthermore, these heaths are now the only place in Britain where all six of this country's native reptiles are found together.

Much of the heathland in Dorset has been lost over the last century and a half. The heathlands that have survived are considerably fragmented, and this has increased the likelihood of loss of heathland and of the biota that the heaths support (Moore, 1962). Haskins (1978) calculated from Isaac Taylor's maps of Hampshire and Dorset that 39 960 ha of heathland existed in Dorset between 1759 and 1765. By 1987 only 5 141 ha of heathland remained (Webb, 1990). It is because so little of the Dorset heathlands remains and because of the rare communities and species that the heathlands support that they are now a top priority for habitat conservation in Britain (Biodiversity Steering Group, 1995).

The fragments which constitute the Dorset heathlands are remnant patches, isolated initially by disturbance of the surrounding area (Forman and Godron, 1986). Important attributes of such fragments include the scale of fragmentation, degree of isolation, fragment size and shape, the species distribution and edge effects (Lord and Norton, 1990).

The Dorset heathlands have undergone both ecological and human-induced change. However, few areas of heathland were actively managed over the period relevant to the present study. Therefore, human-induced change may be treated as a minor source of variation. Most changes in the heathlands are expected to be due to natural processes, such as ecological succession. Ecological forces operate at different spatial and temporal scales, so that change or rates of change may themselves be patchy (Dunn *et al.*, 1991). Therefore, within a single landscape, different patches may experience ecological change at different rates, creating complex patterns of change (di Castri and Hadley, 1988).

The main aim of the research outlined in this chapter was to quantify how variables such as patch size and patch shape may affect the rate of loss of heathland. There are many types of spatial ecological modelling (Czárán and Bartha, 1991), but few models have been developed for ecological processes identifiable at the landscape (patch) scale (for example, Dunning *et al.*, 1992). The integration of ecological models with GIS has much potential (Goodchild, 1993), although integration with object-based GIS is not without its problems (Livingstone and Raper, 1994).

The research was conducted in three stages. First, change in the heathlands was examined on a per-cell basis using the raster data model (and the Arc/Info GIS), as this has been found to be suitable for ecological analysis (Coulson *et al.*, 1991). Secondly, change in individual patches was examined using an object-based approach (Bonham-Carter, 1994). The object-based approach allowed individual cells to be grouped into higher-order objects (patches of heathland). As part of this object-based approach bivariate distribution functions were plotted between percentage change in the area of heath in a patch and several 'explanatory' variables. Finally, a simple model of heathland dynamics was produced using stepwise multiple regression. First, the data are described.

21.2 SAMPLE DATA

21.2.1 The ITE datasets

The Institute for Terrestrial Ecology (ITE) carried out surveys of the entirety of the heathlands of Dorset in 1978, 1987 and again in 1996 as part of a project called 'The Dorset Heathland Survey'. The aim of this project was to examine long-term trends in the heathland landscape. These data have already been the subject of considerable analysis to examine the nature and pattern of change and the effects of area and isolation on heathland biota.

The heathlands were divided into cells of 200 m × 200 m based on the National Grid. Chapman (1975) suggested that a survey square of these dimensions provides a sampling unit that is easily identifiable in the field, yet is sensitive enough to detect changes in the composition of the vegetation. In total, 184 attributes were surveyed which included vegetation types, species of flora and fauna as well as topography and land use types. In 1978, 3 110 cells were surveyed and this increased to over 3 360 in 1987. The data for 1996 were not ready for use. The 184 attributes surveyed were divided into primary and secondary categories. The *nineteen* primary categories were as follows:

1   Dry heath
2   Peatland
3   Wood
4   Grassland

5   Sand dunes
6   Other buildings
7   Wet heath
8   Scrub
9   Carr
10 Agriculture
11 Brackish marsh
12 Farm buildings
13 Houses and gardens
14 Humid heath
15 Bare ground
16 Hedges and boundaries
17 Horticulture
18 Open water
19 Industrial buildings

The nineteen primary categories provided a complete description of each cell and their cover values summed to 100%. The secondary categories gave a more detailed description of each primary category, for example, dry heath is dominated by *Calluna vulgaris* (heather), growing in association with *Erica cinerea* (bell heather), *Agrostis setacea* (bristle bent grass) and either *Ulex minor* (dwarf gorse) or *Ulex gallii* (western gorse). For each cell attributes were scored on a frequency scale as follows: (0) absent, or not detected, (1) present, but less than 10% cover, (2) well represented, but less than 50% cover and (3) dominant vegetation type, with more than 50% cover.

Potential errors in the data sets of this type include those which occurred during the field survey and at the data entry stages. However, precautions taken by ITE during the survey and preparation of the data have minimized errors. Obvious errors (three in all) were identified, the original survey data were examined to find the probable causes and solutions were implemented. However, given the nature of the data (primarily, the coarse numerical resolution of the classes) it is clear that some uncertainty remains.

21.3   ANALYSIS

21.3.1   Estimating areal coverage

The scores obtained from the survey were used to derive estimates of areal coverage for each vegetation type or attribute within each cell. The percentage cover was estimated by adapting an algorithm developed by Chapman *et al.* (1989) in which relative percentage-cover values were assigned to the vegetation types using all the scores for the grid square. The algorithm ensured that the total cover of a square equalled 100%. Once percentage cover was estimated, the total area of coverage in square metres was calculated for each land cover type in each grid square (100% cover was equivalent to 40 000 m²).

21.3.2   Cell-based analysis

Initially, a per-cell approach was taken to examine temporal change in the heathlands of Dorset as a whole. The term heathland refers to dwarf shrub vegetation for the purposes of this study. The coordinates for every square surveyed were entered into the GIS, as

Table 21.1 The total change in area of heathland (square metres) between 1978 and 1987.

Type of heath	Area in 1978	Area in 1987	Areal change	% change
Dry heath	25 872 259	19 924 468	−5 947 791	−23
Wet heath	8 528 297	8 115 838	−412 459	−5
Humid heath	14 792 854	15 892 568	+1 099 714	+7
Peatland	5 915 652	5 852 504	−63 148	−1
Total heathland	55 109 062	49 785 378	−5 323 684	−10

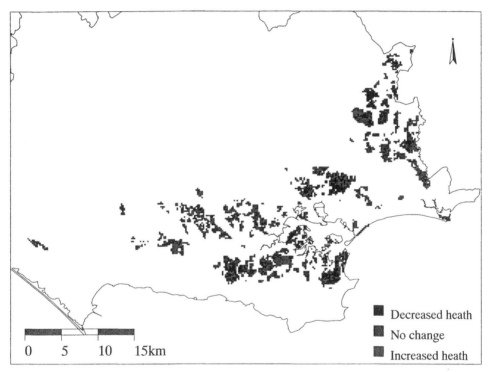

Figure 21.1 Change in the area of heathland on a per-cell basis, 1978–1987. The solid line represents the Dorset County boundary.

was the area of coverage for each attribute to produce a series of grid (raster) coverages. Of the nineteen primary categories, dry heath, wet heath, humid heath and peatland were selected for further study as these comprise heathland vegetation. The grids of the areas of coverage of dry heath, wet heath, humid heath and peatland in 1978 were subtracted from those for 1987 to estimate change on a per-cell basis (Table 21.1). In addition, each heath type was aggregated to estimate the total area of heathland in 1978 and 1987, and the change in area of heathland between the two surveys was estimated by subtraction (Figure 21.1).

The area of dry heath, wet heath and peatland decreased between 1978 and 1987, but the area of humid heath increased by 7%. This percentage increase is thought to be due to severe heath fires caused by the drought of 1976 (Bullock and Webb, 1995), since the fires made accurate surveying of heath type difficult. Overall, however, the total area of heathland in Dorset decreased by 10%. This overall decrease occurred despite the considerable increase in the area of humid heath and the felling of some forestry plantations. These figures differ from those of Webb (1990) because we looked solely at the primary categories, and because our definition of heath (dwarf shrub vegetation) is more restricted.

21.3.3   Patch-based analysis

The cell-based approach allowed an overview of heathland dynamics between 1978 and 1987 (Table 21.1). However, these summary statistics disguise the variation within and between heathland patches. For example, the cell-based approach does not allow one to determine whether small patches are more or less susceptible to change than larger ones. Heathland managers generally require information about patches rather than cells of heath, as the patch is the normal unit of management. Therefore, an object-based approach was adopted to group both the individual cells and their attributes (in this case total heathland) into higher-order objects (patches of heathland).

The patches of heath were created using a nearest neighbour index in the GIS. Each cell of heathland was examined in turn. If a cell containing the same attribute (total heathland) existed either above, below or on either side of the selected cell (including diagonals) it was combined with the original cell to form a patch of heathland, each patch having a unique identifier. In total 116 patches of heathland were created in this way in 1978 (Figure 21.2). This is a different approach to that taken by Chapman *et al.* (1989)

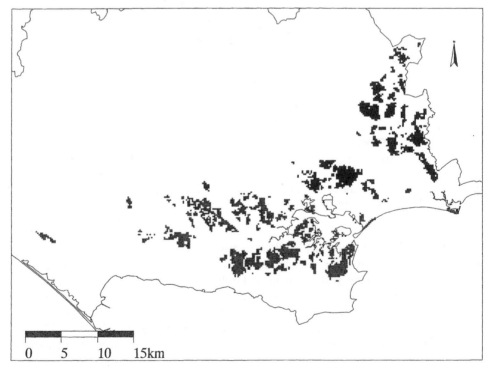

Figure 21.2   Patches of heathland in 1978. The solid line represents the Dorset County boundary.

who joined together cells on the diagonal only if the cells contained over 75% heath. When the survey was carried out in 1978, only cells containing heathland were surveyed. Therefore, all the surveyed cells were contained within the set of patches. We retained this set of patches to analyze the 1987 data and, therefore, discarded the cells surveyed for the first time in 1987, since only decreases in area due to succession and invasion are the present interest.

Once the patches were identified, the area of heathland in a patch was chosen as an indicator of patch status. The area of heathland is not the same as the area of the patch (which is measured in complete cells) and, further, the area of heathland includes only the dwarf shrub vegetation. The response variable was then defined as the percentage change in area of heathland between 1978 and 1987.

The simple difference in area of heathland between 1978 and 1987 might have been plotted on the ordinate in place of the percentage change. The rationale for selecting the percentage change is as follows: if we assume succession by invasion from the edge of the patch, and the rate of invasion is constant, larger patches will decrease in area more than smaller patches. However, if the initial area covered by several smaller patches were equal to the area covered by one larger patch, the area lost would be greater for the smaller patches. The percentage change value embodies the latter logic.

Five explanatory variables (patch indices) were then defined for the patches in 1978 as follows:

1   Area of heathland.
2   Area of patch.
3   Density (area / $c \times 40\,000$).
4   Perimeter.
5   Shape (($\_c \times 40\,000$) / perimeter).

where, c = number of cells.

Numerous alternative explanatory variables could have been chosen. However, the selected variables were chosen because they were deemed of most ecological relevance in a patchy environment, and because it was hypothesized that each would have an effect on the percentage change in the area of heathland in a patch.

First, area of heathland in a patch was chosen (as an explanatory variable) because we expected that the larger the area of heathland in a patch, the smaller would be the decrease in area as a percentage of the initial area. While this is in part a geometrical relation, it follows a model of succession in which patches are invaded from the edge inwards. The area of a patch was included for similar reasons. Secondly, density of heath in a patch was chosen to test the hypothesis that the greater the density of heathland within a patch, the smaller the percentage change in heathland area. In part, this follows a model of succession in which patches are 'invaded from within' with plants establishing from, for example, scrub within the patch, as well as being invaded from the edge inwards. Thirdly, perimeter was chosen because it was expected that, with the initial area of heathland held constant, the rate of succession would be inversely related to the perimeter of a patch: specifically, the longer the perimeter, the greater the percentage decrease in heathland in a patch, as scrub and other vegetation invade from the edge. Shape was chosen as an alternative to perimeter, to test the hypothesis that more rounded (disc-shaped) patches are more robust to change.

Initially, it was necessary to take the logarithm of percentage change, area of heathland, area of a patch, density and perimeter to ensure approximately linear bivariate distributions

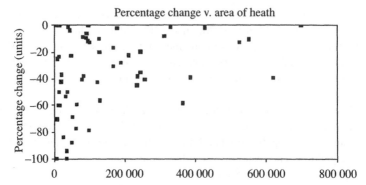

Figure 21.3 Bivariate distribution function between the percentage change in area of heathland and the area of heathland in 1978.

Table 21.2 *F* statistics for simple regression analysis.

Explanatory variable	*F*	Explanatory variable	*F*
Log (area of heath)	21.4	Log (perimeter)	12.4
Log (area of patch)	18.9	Shape	0.36
Log (density)	9.7		

Table 21.3 *F* statistics with log (area of heath) held constant.

Explanatory variable	*F*	Explanatory variable	*F*
Log (area of patch)	0.63	Log (perimeter)	0.016
Log (density)	0.11	Shape	0.64

between percentage change and each of the explanatory variables. Further, it was necessary to remove a small number of patches for which the area of heathland increased, since such increases are likely to be due to some form of management (for example, forest clearance) rather than ecological succession. These increases in area of heath will be investigated at a later date. The bivariate distribution function between the percentage of change in the area of heathland and the area of heath in 1978 is plotted in Figure 21.3.

The bivariate distribution function given in Figure 21.3 illustrates a direct relation between the percentage change and the area of heath. On average, the rate of loss of heathland is greater for smaller patches than for larger patches, thus supporting the hypothesis that smaller patches are more susceptible to invasion.

The results of simple regression are shown in Table 21.2. Where the *F* statistic is equal to, or greater than 4, the variables are significantly correlated at the 95% confidence level. Individually, all of the variables, except shape, were significantly correlated with percentage change (Table 21.2). The logarithm of area of heath was held constant and the other explanatory variables were entered into the model (Table 21.3). However, no further explanatory variables were significantly correlated with percentage change at the 95% confidence level. Thus, the relations shown in Table 21.2 for area of a patch, perimeter and density arise because of a direct correlation with area of heath. That is, it is likely that the explanatory variables influence percentage change through area of heath.

21.4 DISCUSSION

Management practices have an important effect on certain parts of the heathlands. The main aims of the conservation projects in Dorset are primarily to manage the remaining area of heathland, to restore degraded heathland, to expand the heathland area and to ensure its ecological diversity and sustainability. So far these aims are being achieved mainly by the removal of invading species such as gorse, scrub and bracken. However, management does not mean conservation management alone, but rather any human-induced change, be it ploughing or grazing. Management, therefore, can result in decreases in the area of heathland. Although there was little management between 1978 and 1987, data are currently being acquired from the numerous organizations who manage the heathlands, and these data will be used to remove patches of heath which underwent human-induced change between 1978 and 1987. This will not only allow an examination of the consequences of differing management practices on the heaths, but more importantly it will reduce the 'errors' in the per-patch analysis of ecological dynamics.

The heaths of Dorset form a network of patches in a matrix of forest, agricultural and urban land. However, only the heaths have been surveyed. Therefore, data are missing on the matrix within which the heathland patches exist. Webb and Hopkins (1984) state that 'isolated patches of heathland differ from many other habitat islands because they are usually surrounded by communities which are richer in species than the islands themselves and thus, there may be continual pressure from colonisation and succession'. Hence, the nature of the surroundings is an important influence on the heathland patch or fragment. In particular, from an ecological point of view it is important to know if a heathland patch is surrounded by woodland or scrubland which is likely to invade and lead to succession. In these circumstances the patch should be viewed as a component of the landscape in which it occurs. In the future, remotely-sensed Landsat Thematic Mapper (TM) and Multispectral Scanner System (MSS) imagery will be used to map the surroundings of the patches. Veitch et al. (1995) carried out an initial study using the ITE Landcover Map of Great Britain. We will refine this by utilizing the ITE survey data to classify the imagery. The imagery will allow the incorporation of explanatory variables on context.

It is important to remember that the area of heathland in a cell varies, and might be as little as 5%. In part, this explains the different rates of change between cells. When the cells were amalgamated to form patches based on the presence or absence of total heathland, the same is true: heathland was not necessarily the dominant vegetation group in a given patch. Patches, therefore, are themselves patchy, which invites questions about the most appropriate scale of sampling for the analysis. In the future, a 3-D model of species-assemblages in patches will be created to allow a more subtle approach to modelling succession. Species-area curves will be important in this respect (Seagle, 1986). For example, if there is a high density of heathland species around the edge of a patch, does this mean succession is likely to be at a slower rate? Alternatively, if the patch interior contains a high density of heathland species, does this mean the patch will remain intact over a longer time?

21.5 CONCLUSIONS

The per-cell approach provided an important overview. It indicated that the heathlands of Dorset represent a complex spatio-temporal dynamic environment, which may not be in balance. Between 1978 and 1987 the area of heathland decreased by 10%. Further, the heathland did not change in a homogeneous manner: different cells changed at different rates (Figure 21.1).

The main benefit of the object-based approach is that it permitted an examination of the relations between percentage change in heathland area and several variables which describe heathland patches themselves. The bivariate distribution function (Figure 21.3) and regression analyses (Tables 21.2 and 21.3) revealed that heathland change is related predominantly to initial heathland area. This relationship fits our understanding of ecological dynamics in fragmented environments.

There is an urgent need to quantify the problem of fragmentation of the heathlands of Dorset by developing further, more sophisticated, models predicting future change in the heathlands. Such models will be invaluable to the managers of the heathlands. Future research will concentrate on ecological change by incorporating data on management practices into the analysis.

Acknowledgements

The authors wish to thank ITE for access to the database of the Dorset Heathland Survey and thank Dr. Jim Milne and Dr. Dave Martin for their assistance with coding AMLS, and Prof. Anthony Unwin, Dr. Nigel Webb, Mr. Robert Rose, Mr. Ralph Clark and Mr. Kevin Lawless for useful suggestions. The field surveys in 1978 and 1987 were funded by the Manpower Services Commission and BP Petroleum Development Ltd.

References

BIODIVERSITY STEERING GROUP (1995) *Biodiversity: the UK Steering Group Report*, London: HMSO.

BONHAM-CARTER, G.F. (1994) *Geographic Information Systems for Geoscientists, Modelling with GIS*, Computer Methods in the Geosciences, Vol. 13, Oxford: Elsevier.

BULLOCK, J.M. and WEBB, N.R. (1995) Responses to severe fires in heathland mosaics in Southern England, *Biological Conservation*, **72**, 1–8.

CHAPMAN, S.B. (1975) The distribution and composition of hybrid populations of *Erica ciliaris* L. and *Erica tetralix* L. in Dorset, *J. Ecology*, **63**, 809–24.

CHAPMAN, S.B., CLARKE, R.T. and WEBB, N.R (1989) The survey and assessment of the heathlands of Dorset, England, for conservation, *Biological Conservation*, **47**, 137–52.

COULSON, R.N., LOVELADY, C.N., FLAMM, R.O., SPRADLING, S.L. and SAUNDERS, M.C. (1991) Intelligent Geographic Information Systems for natural resource management, in M. Turner, and R. Gardner (Eds), *Quantitative Methods in Landscape Ecology*, New York: Springer.

CZÁRÁN, T. and BARTHA, S. (1992) Spatio-temporal dynamic models of plant populations and communities, *Tree*, **7**, 38–42.

DI CASTRI, F. and HADLEY, M. (1988) Enhancing the credibility of ecology: interacting along and across hierarchical scales, *GeoJournal*, **17**, 5–35.

DUNN, C.P., SHARPE, D.M., GUNTENSPERGEN, G.R., STEARNS, F. and YANG, Z. (1991) Methods for analyzing temporal changes in landscape pattern, in M. Turner, and R. Gardner, (Eds), *Quantitative Methods in Landscape Ecology*, New York: Springer.

DUNNING, J.B., DANIELSON, B.J. and PULLIAM, H.R. (1992) Ecological processes that affect populations in complex landscapes, *Oikos*, **65**, 169–175.

FORMAN, R.T.T. and GODRON, M. (1986) *Landscape Ecology*, New York: Wiley.

GOODCHILD, M.F. (1993) The State of GIS for Environmental Problem Solving, in M. Goodchild, B. Parks and L. Steyaert (Eds), *Environmental Modeling with GIS*, Oxford: Oxford University Press.

HASKINS, L.E. (1978) *The Vegetational History of South-East Dorset*, PhD thesis, Southampton: University of Southampton.

LIVINGSTONE, D. and RAPER, J. (1994) Modelling environmental systems with GIS: theoretical barriers to progress in M. Worboys, (Ed.), *Innovations in GIS 1*, 229–240, London: Taylor & Francis.

LORD, J.M. and NORTON, D.A. (1990) Scale and the spatial concept of fragmentation, *Conservation Biology*, **2**, 197–202.

MOORE, N.W. (1962) The heaths of Dorset and their conservation, *Journal of Ecology*, **50**, 361–391.

SEAGLE, S.W. (1986) Generation of species area curves by a model of animal habitat dynamics, in J. Verner, M. Morrison, and C.J. Ralph (Eds), *Wildlife 2000: Modeling Habitat Relationships of Terrestrial Vertebrates*, Madison, Wisconsin: University of Wisconson Press.

WEBB, N.R. (1990) Changes in the heathlands of Dorset, England, between 1978 and 1987, *Biological Conservation*, **51**, 273–286.

WEBB, N.R. and HASKINS, L.E. (1980) An ecological survey of heathlands in the Poole Basin, Dorset, England, in 1978, *Biological Conservation*, **17**, 281–296.

WEBB, N.R. and HOPKINS, P.J. (1984) Invertebrate density on fragmented *Calluna* heathland, *Journal of Applied Ecology*, **21**, 921–933.

Index

T - #0006 - 071024 - C4 - 254/178/15 [17] - CB - 9780748408108 - Gloss Lamination